GEOPOLITICS OF NATURAL GAS

A REPORT OF HARVARD'S ENERGY AND ENVIRONMENTAL POLICY CENTER

GEOPOLITICS OF NATURAL GAS

JEREMY RUSSELL

BALLINGER PUBLISHING COMPANY
Cambridge, Massachusetts
A Subsidiary of Harper & Row, Publishers, Inc.

International Standard Book Number: 0-88410-610-1

Library of Congress Catalog Card Number: 83-2670

Printed in the United States of America

Library of Congress Cataloging in Publication Data

Russell, Jeremy.
Geopolitics of natural gas.

Bibliography: p.
Includes index.
1. Gas industry—Political aspects—Europe.
2. Gas industry—Political aspects—Soviet Union.
3. Power resources—Europe. 4. Power resources—Soviet Union. 5. Liquefied natural gas industry—Political aspects—Arab countries. 6. Gas, Natural—Arab countries. 7. World politics—1945- . I. Title.
HD9581.E8552R87 1983 333.8'233 83-2670
ISBN 0-88410-610-1

CONTENTS

LIST OF FIGURES

LIST OF TABLES

ACKNOWLEDGMENTS

The opportunity to write this book was provided by an arrangement between Shell International Petroleum Company Ltd. in London and the Energy and Environmental Policy Center at the J.F. Kennedy School of Government in Cambridge, Massachusetts, under which I studied as a Visiting Research Fellow at Harvard between October 1981 and June 1982. My warm thanks are therefore due to both the EEPC and to Shell, although the views expressed in this book are solely my own and do not represent those of either Shell or the EEPC.

I particularly wish to acknowledge the very helpful and constructive comments and suggestions made by Bill Hogan and Henry Lee and the invaluable discussions that I had with Al Alm, Colin Bowkley, Phillip Clendenning, Napier Collyns, John Curtis, Marshall Goldman, Boyce Greer, Michael Nacht, Joseph Nye, Floris Ouwehand, Tom Stauffer, Dan Yergin, and numerous others.

I also wish to thank Malcolm Hamilton, Liz Bernstein, Judith Gurney, Marianne Stevenson, Sharyn Imonti, and others on the staff of the EEPC and the Kennedy School for their valuable assistance on administrative matters. Finally, thanks are also due to Mary Petersen, who was responsible for so much of the preparation of the manuscript, and above all to my family for their transatlantic support and encouragement.

GEOPOLITICS OF NATURAL GAS

1 INTRODUCTION

For many people in Western Europe and North America, the only close encounter they are likely to have with natural gas is in the kitchen or in the domestic water-heater room. They may admire its blue, clean flame or worry about the smell caused by the chemicals that are added to the otherwise odorless methane to reveal when it is escaping from its metal prison. They may mutter to themselves about the cost of the gas they buy and may try to evaluate it against other available fuels suitable for achieving their objectives of cooking and heating. But how many people stop to think about how they come to be using natural gas at all? How many care where it comes from as long as it keeps coming? There are probably few people who know much about prospects for the future supply of natural gas from their own country's resources; probably fewer still—except perhaps in West Germany, France, and Italy—are interested in what role, if any foreign countries may already be playing in enabling the clean blue flame to keep on burning in their homes or how the future of their domestic energy usage can be profoundly affected by happenings in Norway or North Africa or by deliberations in Moscow or Ottawa. In countries where gas of one sort or another has been used for many decades, it seems to be taken for granted. Rather like water, it is available at the turn of a tap and thus, it may never be truly appreciated until it is suddenly no longer available. So what are some of the international elements that are, or may become, involved in influencing the supply of natural gas from where it is found to where it is needed?

This book aims to illuminate some of the complex issues of internationally traded natural gas and to see what may need to be done to ensure that, in a world faced with the problems of providing for its increasing energy needs against a background of quickening changes in the relationships between nations, users and producers of this important global energy resource can continue to benefit from its existence. It does not attempt to cover all aspects of either the gas industry or international politics, and it does not set out to be a technical or statistical handbook on natural gas, although, where it is considered essential for an understanding of the broader issues, some factual data are included. During the preparation of this book the author had recourse to many more sources of information and shades of opinion—some published, some not—than it would be appropriate or indeed possible to itemize fully as footnotes in the following pages except in specific cases. However, a selected bibliography is included, and it will provide those interested in further examining the issue with a solid and helpful basis of reading material.

In view of the turbulent times in which we live, it would be overly ambitious to make predictions about how the international trade in natural gas will shape up between now and the end of the century. It is possible, however, to distinguish certain trends both in the way the world is approaching its energy problems and in the way its nations are handling their relationships with one another. These trends can guide policymakers responsible for the efficient mobilization of the world's natural gas resources.

The reader is encouraged, therefore, to accompany the author on a journey of investigation into the role that natural gas now plays in meeting the world's energy requirements and that it may be able to play over the next two decades, if certain things are done by certain countries in certain ways and at the correct times. This book will concentrate upon the West European scene, partly because of Europe's geographical location in relation to the main gas-producing areas and partly because of its traditional political orientation and involvement in international issues. The relevance of both these factors is evident today in the controversy over the future of natural gas supplies from the Soviet Union, and this issue will be dealt with at some length. But the countries of Western Europe have, as a result both of their experience of the exercise of power on a global scale and of their none too generous allocation of hydrocarbon resources, historically played a major role in the evolution of international energy movements. The current degree of dependence upon imported energy experienced by most West Europe nations is likely to mean that they will continue to have more than average interest in the successful management of international energy issues. This is not to say that other major energy-importing nations around the world do not also have major interests or roles to play in that management—Japan and the United States are obvious examples of countries that do—but there is a certain element of geographical isolation involved with both of them. While their rather special cases will be discussed, it is not felt that they are quite such complex cases as is Western Europe.

The shocks to the international energy system delivered in the 1970s resulted in such hefty rises in the price of internationally traded oil that all importing countries were obliged to reexamine their energy consumption and supply arrangements and to rethink their plans for the development of the different energy resources available to them. While some governments sought to shield their people from the full impact of the price rises by imposing controls on domestically produced energy, others allowed the ultimate consumers to bear the full brunt. This inevitably initiated a process of change in consumer habits that has by no means run its full course, and everywhere people have been alerted to the need to conserve energy. Governments are embarked upon the difficult and tortuous process of creating policies that will optimize energy supply utilization in such a way as to minimize the adverse economic and political repercussions stemming from a protracted period of constraint in fossil fuel availability, which seems likely, in spite of occasional periods of temporary oversupply, to last well into the next century.

Natural gas currently supplies slightly less than one-fifth of the energy consumed worldwide, and its consumption is concentrated in North America, the Soviet Union, Western Europe, and Japan. With the exception of Japan, which has to import virtually all its energy, the development of markets for natural gas has been based upon substantial indigenous reserves although the evolution of those markets has been quite different in each of the areas. Major undeveloped reserves of natural gas remain in the Soviet Union, whereas both Western European and United States resources have probably passed peak production capability. A great deal of the remaining discovered global resources are found in the Middle East, and in North and West Africa, where little market exists at the current time. In the early days of hydrocarbon exploration, natural gas was definitely not what people were looking for, and when it was discovered in wells drilled with the express purpose of finding oil, it was even considered by the oilmen as something of a nuisance. Vast quantities of natural gas were flared—that is, burnt off in the atmosphere—or otherwise wasted, and even today, particularly in the Middle East and West Africa, flaring still continues where natural gas is found associated with crude oil. There are few alternative outlets, and the construction of albeit costly gathering and treating facilities that would allow associated gas to be harnessed for domestic consumption or export schemes is sorely needed.

Because natural gas is a low-density energy source when compared with oil or coal, it is much more expensive to transport and distribute, and its simple chemical composition precludes it, for the time being at any rate, from finding uses in many of the outlets in the transportation and chemical manufacturing fields in which its two major competitors are so indispensable. There are those who consider that these limitations on flexibility of use are probably destined to prevent natural gas from ever playing as significant a role in meeting global energy requirements as its geological resources, which now approach those for oil in terms of energy equivalence, indicate that it could. On the other hand, natural gas has certain characteristics—e.g., cleanliness of combustion and an ability to be used directly as a primary energy source in the home, without elaborate processing or manufacture—that make it particularly attractive as a domestic fuel and in certain industrial applications where purity control is important.

The troubles afflicting the international oil industry in recent years have done much to promote the cause of internationally traded natural gas, and one of the purposes of this book is to examine whether the political influences that, in combination with others, have given rise to considerable unease on the part of the major consuming countries about the future security of energy supplies, will play a similar role in respect of natural gas as they have played and are still likely to play in respect of oil.

One of the conclusions arising out of the study that generated this book is that there are some good reasons why the future of internationally traded natural gas may not be quite so bedeviled by politics as oil has been, why there is less likelihood of any organization similar to OPEC evolving in the gas trade, and why there is every chance that gas will maintain and perhaps slowly increase its share of the world's primary energy supply. This is not to suggest that it will necessarily do so in Western Europe or that the future of Western Europe's natural gas supply will be by any means free of shocks caused by political manipulation. But the future of imported coal and uranium, let alone the development of nuclear electricity in Western Europe, seems far from trouble free today, and no country or group of countries that has to rely on substantial energy imports should expect to find an energy supply source that is immune to such trouble.

In some West European countries there are market sectors—e.g., public distribution outlets in the Netherlands—that may be close to saturation in respect of natural gas. Other sectors may use increasingly less gas, either as a result of government policies or because of the growing availability of competitive fuels. But, in a few European countries, the public distribution sectors will actually expand, and pressures for reduced natural gas consumption generally will be somewhat offset by prospects for increasing consumption in Italy and France and in those countries such as Spain, Sweden, and Denmark where natural gas has so far played a relatively minor role in the fuels/energy balance and where there is great potential for expansion.

Another conclusion is that Western Europe, in view of its declining indigenous production but static or slowly growing demand, is going to increase its dependency upon imported gas. In the West European countries that have traditionally used gas, there is a widespread but apparently inflexible infrastructure into which has gone a great deal of investment over the decades. Even if it were not enlarged, this infrastructure has a lifetime extending into the next century, and it is too valuable to be allowed to be underutilized. However, the prospect in West Germany, France, Italy, and Austria, for example, is for static or declining domestic natural gas production from now on and, as will be shown, this must mean substantially increased imports.

Not surprisingly, price considerations will play an extremely important part in determining just how much the demand for natural gas will increase or decline. While the former trend may be indicated if gas is available at prices at which it can compete with fuel oil in the marketplace—that is, at or near fuel oil parity—the latter trend is more likely if the price rises to near or above crude oil parity. This development is currently being strongly urged by certain producer countries, particularly where these have been exporting crude oil in the past and may already be faced with dwindling oil export capabilities.

Western Europe is geographically conveniently placed to receive natural gas imports not only by pipe but also by sea in the form of liquefied natural gas (LNG) from two of the world's major resource-rich areas: the Soviet Union and North and West Africa. Unless free access via the Suez Canal is available to LNG tankers, it may be too far from the Middle East for many commercially viable export projects to be created before the end of the century. But imports into Western Europe from this part of the world are not likely to be particularly relevant to the long-term objective of enhancing energy security generally by diversifying sources away from, for example, OPEC oil. So long as the United States remains an unpromising market for imported LNG, certain African producers have little option but to seek outlets for their surplus natural gas in Western Europe.

In addition to the more commercial considerations, this book attempts to illuminate some of the political factors likely to influence the global trade in natural gas. While many people would consider the East-West relationship the most important single factor that has to be taken into consideration when looking at international issues, there are others, such as the North-South dialogue and growing economic nationalism, that may become even more important, assuming that the basis upon which international commercial, political, and diplomatic relations are to be carried out over the next twenty years remains more or less as it currently is. Perhaps this is too bold an assumption at a time when the possibility of some major East-West military confrontation is increasingly openly talked about, and not only by politicians and the media. This book is based on the assumption that major isolationist/protectionist trends will not develop and that economic growth in the industrialized countries of the West will pick up over the next ten years, and with it, a sustained growth in international trade. It also assumes that the international financial system holds up under the increasing pressure it now experiences and that longer term credits and other forms of financing will be available in sufficient quantities to allow energy-related projects in general and natural gas projects in particular to be initiated and carried out effectively and efficiently.

CHANGES IN THE INTERNATIONAL SYSTEM

The period of comparative international stability that obtained after World War II was largely financed and policed by the United States, which alone had the economic and military might to dominate any world system and particularly one that it was primarily involved in creating. During the last thirty years, however, the international power balance has changed markedly, partly as a result of increasing Soviet military and economic capabilities but more importantly as a result of a diffusion of power away from the former imperial and colonial nations to a new group of nations anxious to exert their newly found independence. The usable power of the United States itself began to decline in relative terms as first Western Europe and then Japan recovered from the war, with U.S. help, and started to compete with her economically. At the same time, the Soviet Union, on the back of an anti-imperialist surge in global sentiment and as a result of a steady and determined effort to build up its military forces, increased its ability to influence global events, not so much for its own direct material or ideological benefit as for the disadvantage of the "West" as it had become loosely termed.

The ability of the West to get as much out of the global system as it once was accustomed to getting declined, and while some Western nations accepted this as inevitable and sought, however ineffectually, to adapt to the changed situation, others were slower to appreciate what was happening. They found themselves opposing the changes and indeed attempting to restore the status quo to something they were both used to and used to controlling. In short, control of the international relations order gradually passed from the great powers of the West and, while it never passed to the "East," usable power became so widely distributed that, by the early 1980s, control could hardly be said to be exerted by anybody except, to a limited degree, locally. A number of new regional power centers did indeed develop, and certain of the newly industrialized countries of the Pacific Basin became so competitive with the old industrialized countries of the West in traditional sectors of their economies that, when the energy shocks occurred in the 1970s, the latter were in poor shape to resist or stave off the ensuing decline in their economic global standing. Subsequently, even growth in their standards of living seemed to be threatened, and this was by the early 1980s already presaging widespread societal changes.

As more and more nations became, as they perceived it, masters of their own fates, nationalism emerged as the main force driving the international political engine. Nowhere was this more strongly manifested than in the economic field, and economic nationalism proliferated, particuarly where the exploitation of natural resources and energy was concerned. New attitudes and new groupings developed within and between nations. This assisted in the reduction of stability and predictability in economic as well as other relations, and gave rise, in a few countries, to feelings of intense frustration and anger. It also gave rise to the need to find explanations, apart from national behavior, for any adverse developments.

The ideologically motivated outpourings of the Soviet Union, coupled with the undeniably worrying increase in Soviet military strength, provided an excellent target for certain Western frustrations. But, although the subsequent concentration of so much political, military, and economic effort on combating what could only be seen by the West as Soviet expansionism may have been justified and may still be, it has perhaps tended in certain countries to become almost an obsession, blinding their politicians not only to other significant geopolitical developments, such as nationalism and religious fundamentalist revivalism, but also to possibilities for adaptation to such developments. There were good reasons, in mid-1982, for expecting preoccupation with the East-West confrontation to continue for some time, although it was not impossible that economic pressures facing both superpowers might soon oblige them to take steps to at least moderate their belligerency and curb their inceasing military expenditures.

Meanwhile, it was also possible to detect a trend in some of the developing countries toward a degree of economic pragmatism that recognized that political independence and economic interdependence are not easy to separate and that economic nationalism may not always be sufficient to bring about the satisfaction of rising economic expectations. This applied particularly to countries whose embryonic infrastructural and technological abilities were confronted for the first time with the harsh realities of competition in the global marketplace and with the problems associated with rapid population increases and rising energy imports costs.

Turning back to the natural gas scene, only a relatively few countries that possess substantial reserves of natural gas are likely to be in a position to compete for the West European markets, with or without technical and financial cooperation from outside. They range from small, developing countries with little or no previous experience of the natural gas industry, let alone the international trade in natural gas, through clearly Western-oriented industrialized countries with established domestic and export gas industries, to the Soviet Union itself, the largest natural gas exporter in the world, but a country that is far from approaching any reserve-associated limit to its export potential. As might be expected, the political interests of all the producer countries concerned were not, in 1982, particularly closely identified and seem unlikely to become so. While their economic interests in exporting natural gas could be similar in general, their particular domestic economic positions differed widely and, in respect of the West European import market, they were finding their interests conflicting strongly in a number of cases.

In reality there is no such thing as "the West European market" for natural gas since, although the gas-importing countries of Western Europe are in fairly close geographical proximity to one another, their interests in importing natural gas from outside Western Europe vary from none at all to substantial, and this is particularly relevant when imports from the Soviet are under discussion. While several countries had, in 1982, already been importing Soviet natural gas for a number of years, and seemed likely to import increasing quantities in the future, others were not expected to do so, or if they should, then only in relatively small quantities. Those countries already taking Soviet gas tended to regard the prospect for increased deliveries with equanimity and as merely a routine extension of a mutually beneficial form of economic cooperation. Newcomers to Soviet gas see it as a convenient alternative source of hydrcarbons that enables them to diversify away from imported Middle Eastern oil. Indeed, most of the controversy raging in the early 1980s over the so-called Yamburg pipeline deal was generated by countries that were not directly involved but feared that the Western Alliance could be politically harmed by the growing economic interdependency that was clearly implied by any increase in energy-import dependency of the West upon the East.

The Soviet Union is known to have a substantial economic stake in the expansion of its natural gas exports, as it faces dwindling oil exports and declining hard-currency earnings. Furthermore, nearly all Soviet gas to date has been exported in exchange for imports of large-diameter pipeline, compressor stations, and related equipment needed for the development of the domestic gas network. By the end of the 1970s, with coal production faltering and oil production barely increasing, natural gas was scheduled to play a more important role in the fuels/energy balance of the Soviet Union for the remainder of the twentieth century, and by making larger quantities available for export, the Soviet Union hoped to be able to afford to import the hardware necessary for expanding its internal natural gas infrastructure. Since an ongoing and growing Soviet requirement for hard-currency imports is likely for the foreseeable future, those in favor of an expansion of Soviet gas exports to Western Europe claim that mutual interest in the continuing flow of the gas is such as to reduce to zero any increase in unilateral dependency of the West upon the East. The Soviet Union, it is argued, cannot at the same time be a reliable supplier of natural gas to western Europe—justifying the extension of long-term large scale credits with which to purchase large quantities of Western equipment and technology—and play political games by threatening to turn off the gas taps. This subject is treated at some length later in the book. Even if the original Yamburg scheme is greatly modified, or even "killed" for the time being, there seems a lot of compelling geographical logic on the side of those working for larger East-to-West flows of natural gas on the European continent. It may be that less attention should be paid to the East-West confrontational aspects than has been the case in the early 1980s.

One beneficial result from all the attention that was focused on the Yamburg pipeline deal was that the question of energy supply and security was given a grand and much-needed airing, and few people in government or industry can now be unaware of the problems involved, both generally and in respect of natural gas. The real meanings of energy dependency, sectoral dependency, interfuel substitution, risk spreading, stockpiling, allocation schemes, and the like are being hotly debated and more sharply defined. Enhanced cooperation among West European countries most closely involved, to reduce the potential damage to economies and to the well-being of individual consumers, has not only been talked about but, in some cases, actually acted upon. It is now to be hoped that if Western European countries have little alternative but to take large quantities of Soviet and other imported natural gas over the next twenty years, then at least they will be well prepared to manage any situation involving interruption of supplies. Those who argued vehemently that "Western Europe" would become more susceptible to Soviet political manipulation as a result of the energy deals had, in late 1982, yet to demonstrate convincingly what they actually meant and how this would come about in practice. But a number of major issues were raised by the very existence of the controversy, for example: a possible transatlantic divergence of interests; the future and objectives of the Western Alliance; the evolution of a united Europe; and the wisdom of attempting to isolate a military superpower, particularly one that was largely self-sufficient for nearly all strategic raw materials including energy but with the possible exception of grain.

By the end of this journey of investigation, it is hoped that the reader will be in a clearer position to view the role that natural gas can and may play in meeting Western Europe's energy requirements over the next two decades. Numerical predictions, where they are made, serve only as indicators of the order of magnitude likely to be involved. There is no greater likelihood that the figures will turn out to be correct than that the political trends examined herein will give rise to clear-cut changes on the international political scene.

Throughout this book, natural gas is taken to mean a mixture of gases from underground accumulations in which methane (CH_4) predominates and, except where otherwise stated, has an energy equivalent of 9,500 Kcal/m^3 = 1,000 Btu/ft^3. When the word "billion" is used, it means the same as the European milliard (mrd) or 10^9, and trillion is 10^{12}. One mrd m^3 or one billion cubic meters (bcm) of natural gas is approximately equal to 890,000 tons of oil. LNG is natural gas cooled to minus 161 degrees centigrade at atmospheric pressure, and 1 million tons of LNG is equivalent, to 1.4 mrd m^3 of gas, volumetrically, 1.2 million tons of oil, or 1.9 million tons of hard coal.

2 THE GLOBAL ENERGY SCENE

It would be far too ambitious in a work of this nature to attempt to cover, even superficially, all the many and complex aspects of the global energy situation, but there have clearly been developments in recent years, an understanding of which is necessary if a clearer perception of the role that natural gas can play in meeting the future energy requirements of Western Europe is to be achieved. These developments can perhaps be described simply as man-made limitations on the availability of global oil supplies. They led, on the one hand, to massive increases in the prices of internationally traded energy, and on the other, to a greater realization of the actual natural limitations on conventional, nonrenewable energy forms, given the finite nature of fossil fuel resources and the extremely rapid increase over the past two hundred years in the rate at which they are being used up. The so-called oil shocks of the 1970s demonstrated the degree to which industrialized countries had become dependent upon imports of oil to meet their growing energy requirements. They also served to bring home to people, in perhaps the most effective way possible, the fact that there had to be some moratorium on energy consumption increases generally to allow both the tapping of new energy resources and the development of new technologies that would increase energy use efficiency and energy saving. Such developments could one day possibly allow the decoupling of economic or industrial progress from a comparable increase in energy consumption, the situation to which people had become accustomed in recent decades. It was possible in 1982 to feel relieved that the oil shocks came when they did and not at some future date, when it would surely have proved even harder to extricate the global economic system from the resulting turmoil than it is now proving.

Some people were aware of the inevitability of some sort of energy crisis even before the shocks of the 1970s, but their voices received little attention amid the clamor and bustle of rising living standards, faster transportation, and rapid industrial expansion. So long as the oil and gas kept flowing, not much attention was directed toward doing much to cope with the situation that would arise if and when they stopped. The prospect of cheap electricity from nuclear fission was alluring, and the possibility that one day there might even be inexhaustible power from fusion enabled substantial research programs in both fields to be supported both intellectually and financially, although, perhaps inevitably, all too little effort was devoted to the practical and political problems associated with widespread dissemination of the tangible technological results. While energy from fusion is not likely to become available on a significant scale until well into the next century, fission-based nuclear energy is beginning to make an impact on electricity availability in a number of, mainly, fully industrialized countries. But even here there are formidable technical, political, and environmental barriers to be overcome before it can even approach the potential contribution to meeting energy supply requirements of which it is capable. Again, it will not be before the end of this century that these problems will be satisfactorily solved to the point where many developing countries can directly benefit from the availability of domestically produced nuclear power.

It needed something more than the pronouncement of wise men to get home to people that they ought not to be using energy in as profligate a manner as that to which they had become accustomed, and for quite a while, even after the first oil shock, there was widespread refusal to believe that this was anything more than a temporary and anomalous condition. There were indeed many who were prepared to ascribe the whole thing to some insidious plot between the oil producers and the multinational oil companies to drive up prices. The short-sightedness of such people could well have hindered or delayed the adoption of many measures that might, by the time of the second oil shock, have gone a long way to alleviating some of the difficulties that ensued. But, in spite of all the evidence, it took that second shock to achieve what the first one was only partially able to —namely, a fundamental change in the attitudes of contemporary consumers toward energy, consequent upon an awareness of just what the implications for future generations of consumers might be.

Even today it would probably be incorrect to speak of the global energy situation as being in crisis, in that there is a great deal of fossil fuel as yet untapped, but it is by no means all in the form in which it is most useful, and access to much of it is increasingly expensive and either geographically inconvenient or technically and commercially difficult. In the short term, there is usually perceived to be a conflict of interests between oil producers and consumers, which has caused the latter to examine ways in which they can reduce their dependency upon the former. In some cases certain oil-importing countries attempt to spread their risks by seeking supplies from nontraditional sources, but in many others it has become a question of reducing the share of total energy supply contributed by oil from whatever source.

The main problem with many of the schemes for diversifying away from oil and into other energy sources is the long lead time involved, for financial, technical, environmental, bureaucratic, or political reasons, in getting them off the ground. As it has taken a long time, over half a century, for the current degree of oil dependency to evolve, so it will take a long time, but hopefully only a few decades, for that dependency to be reduced, albeit by increasing dependency upon other energy sources, conventional or new. But a great deal can happen during a decade that can profoundly affect the capability and the political or social will of a nation to undertake and carry to completion the changes, both in people's mindsets, from reliance on an old energy source toward reliance on a new one, and in the domestic and international infrastructures that would be necessary. It is not just a question of changing one infrastructure for another, but of creating a new or expanding an old infrastructure at the same time that the one used for the supply of the less favored fuel remains partially but decreasingly in use.

National governments have a variety of fiscal and legal instruments at their disposal for ensuring that the changes that they have proposed are brought about. In those countries where the market system is still given free rein, however, and where government intervention in the operation of the market is disfavored, it is often left to individual domestic and industrial consumers to make their own decisions about what type of fuel they will use. This decision is often based on what their pocketbooks tell them. It complicated things for example, in 1981, when everyone was fired up to change from oil to natural gas, electricity, solar, or coal, that there should appear to be developing a "glut" of oil such that the high prices once enforced by the producers became no longer tenable and oil became more readily available once again. The significance of lead times in this connection was that, while the so-called oil glut was clearly a short-lived phenomenon, investment in alternative energy forms that would not come on stream (i.e., start contributing to the overall supply) for at least eight or ten years had to be made at a time when energy costs appeared to be falling below the level that might be considered commercially attractive for such investment. Awareness and understanding of the nature of the energy problems is one thing; the technical and industrial means of solving them is another. But a stable and clear-cut financial environment, likely to obtain throughout the lifetime of the investment in a new energy form, is something else again. Much foresight, faith, and preparedness to take risks will be necessary if by, say, the end of the 1980s or early 1990s, the world's main energy consumers are to avoid an even more serious set of problems than those they currently face.

In this connection it should be remembered that, while the developed or newly industrializing countries have been the major consumers of global energy supplies and have undoubtedly been hard hit by the rise in global energy prices, even to the point where they may find it increasingly difficult to extricate themselves from the grip of recession, it is among the oil-importing developing countries that the greatest financial problems have to be faced. In these countries, energy consumption per capita is very low, and yet it seems that only through a substantial increase in energy consumption in one form or another can they make any significant industrial or agricultural progress. Not only are traditional forms of domestically produced energy, such as firewood or animal dung, becoming scarcer or required for other, more vital, purposes, but also liquid fuels, based on imported oil, are apparently the most convenient and appropriate energy forms, given the existing distribution capabilities and the great necessity to improve transportation infrastructure. Furthermore, in many developing countries, the intellectual and technical resources are insufficient to cope with the infrastructural changes involved in handling the switch away from dependency on imported oil, the cost of which is absorbing an increasing share of the foreign currency earnings of these countries.

Developed countries, for their part, are in the process of substantially reducing the quantities of oil they need to import, as a result of conservation measures, increased energy-use efficiency and, in some cases, increased domestic production of alternative energy sources. But many of these options are scarcely open to many developing countries, even though the latter can usually secure investment finance for such endeavors as domestic oil and gas production or exploration projects. The requirements of developing countries for imported oil are destined to grow for the foreseeable future even if massive assistance in the development of indigenous energy reserves from international bodies such as the World Bank is immediately forthcoming. But the ability of many of these countries to borrow money simply to pay for energy imports must be somewhat limited, and already a troublingly high proportion of their national income is committed to the servicing of their substantial indebtedness. Clearly, they have been faced with severe problems arising out of high global energy prices, but their rapidly growing populations militate against their finding early solutions, with or without technical and financial aid from the developed countries.

A global awareness of impending energy supply problems has, as we have seen, perforce been created by the rather special case of the international oil situation that developed during the 1970s. Hopefully, the world is now embarked, with varying degrees of reluctance and sense of purpose, upon a transition period away from such heavy dependence upon nonrenewable fossil fuels toward greater reliance upon renewable or inexhaustible energy sources. But while there are good prospects that many of the technical problems associated with that transition will be solved by the middle of the next century, there is no escaping the fact that conventional fossil fuels, including nuclear energy based on uranium ore, will still provide the major part of global energy requirements long after the year 2000, which is the horizon set for this book. How well individual countries handle that transition depends not only upon technological capabilities but also upon the way the international political and financial systems are operated in the light of all the other pressing global problems, such as overpopulation, environmental pollution, the provision and distribution of food and drinkable water, the reduction in the incidence of disease, the control of the trade in armaments, and the proliferation of nuclear weapons. But the geographical maldistribution of fossil energy reserves in relation to the main energy-consuming areas will remain as a fundamental consideration in the future development of the international energy trade and will provide numerous opportunities for both international cooperation and geopolitical machinations.

The global energy situation of the early 1980s is thus marked by acute sensitivity and awareness on the part of fossil fuel producers and consumers alike that change and transition are underway and that, given the basic instability of the international political situation, there is every likelihood that transition will not be effected in a smooth and orderly manner and that numerous shocks to the energy system may well be in store. But, while it would be futile in such circumstances to attempt to predict how things will turn out over the next twenty years, it is possible, by means of a scenario approach, at least to discern some of the most likely developments and to get an idea of what is or is not plausible. It is appropriate to establish some sort of basis upon which to develop such scenarios in energy and related fields and, while a great many scenarios could be constructed, three will be enough to illustrate the general directions that events could take. These will be described later. If, however, one first examines the basic situation with regard to oil supply and demand, it becomes apparent that, while there has been little change over the past few years in estimates for ultimately recoverable oil reserves, there has been a substantial reduction in oil demand forecasts as a result not only of conservation and substitution but also of the lower economic growth now expected in the main oil-consuming countries of the Organization for Economic Cooperation and Development (OECD). The effect has been to defer the consumption of some 150 billion barrels of oil into the next century, so that the decline in global oil production that was confidently forecast, until quite recently, to occur over the next two decades is postponed for fifteen or twenty years. It looks now as though even if crude oil demand in the noncommunist world should climb above, say, 55 million barrels per day by the year 2000, such a production level could be maintained until at least 2010, with a slow decline thereafter. This picture would of course be further improved if demand for crude oil could be reduced even more.

The OECD countries are, for the most part, postgrowth in energy consumption and are in many cases currently experiencing sharp reductions in primary energy use per unit of Gross Domestic Product (GDP). Under any of the three scenarios that will be described later, it is likely that there will actually be a drop in oil demand of as much as 10 percent between now and the end of the century, so that OECD demand in the year 2000 would probably not exceed 35 million barrels per day and might even be as low as 27 million barrels per day. Declining oil demand is expected to be seen most spectacularly in Western Europe, where there could be a drop of around one-third from the peak levels reached in 1979. This contrasts with the substantial increase expected in both energy and oil demand in the developing world as a result not only of population growth and industrialization but also of the growing urbanization now clearly underway. Oil demand in these countries could rise by some 15 million barrels per day to the 24 to 27 million barrels per day range by the year 2000. It is clear that the OPEC countries themselves will be using a lot more oil and that this could account for some 6 to 7 million barrels per day, but the remaining incremental 10 million barrels per day will need to be imported by the non-OPEC developing countries. (See Table 2–1.)

In 1980, global primary energy demand was 134.9 million barrels per day of oil equivalent, of which North America consumed about 42 million barrels per day, Japan 7.3, and the Communist countries about 41 million. A reasonable range for the year 2000 could show a global figure of between 197 and 218 million barrels per day, of which North America might consume between 50 and 54 million, Japan 10 to 11 million, and the Communist countries 65 to 68 million. In 1980, the global energy demand of 134.9 million barrels per day of oil equivalent was met by 61.6 barrels per day of oil, 25.3 of gas, 34.8 of coal, 3.3 of nuclear, and 9.9 million barrels per day of oil equivalent by hydroelectric and other energy forms. If the range of 197 to 218 million barrels per day of oil equivalent taken for the year 2000 is broken down in a similar way, oil might supply

Table 2-1. Oil Demand (million barrels per day).

	1980	*2000 (Estimate)*
OECD	34.9	27-32
Developing Countries	10.7	24–27
—of which OPEC requires	2.8	6- 7.5

Table 2-2. Global Energy Demand by Areas (million barrels per day oil equivalent).

Areas	*1980*	*2000 (Estimate)*
North America	42.2	50– 54
OECD Europe	25.5	27– 30
Japan	7.3	10– 11
OPEC	4.1	11– 15
Communist areas	41.2	65– 68
Rest of the world	14.6	34– 40
	134.9	197–218

between 70 and 73 million barrels per day, gas between 41 and 45, coal between 58 and 60, nuclear between 14 and 18, and hydroelectricity and other forms between 17 and 20. (See Tables 2–2 and 2–3.)

There are, of course, a great many factors that must be taken into consideration when working up even comparatively broad-ranging forward estimates as the ones given above, and economic growth rates are clearly among the most important of these. Not only are there wide ranges possible for the overall global picture, but there are wide geographical and country variations and variations within the economic groupings of the OECD countries, the newly industrialized countries, and the developing countries. The most that can be attempted in this chapter is a general indication of the direction things seem to be headed, barring any major "accidents." The economic outlook in the short term, to the mid-1980s, remains highly uncertain wherever one looks, with the economies of the OECD countries continuing to underperform, with industrial production activity leveled off at a low plateau, and with little to indicate when the long-awaited recovery might begin. This underperformance is characterized by a reduction in growth potential, a weakness in investment due to business uncertainty about future levels of demand and inflation coupled with a general perception of a continuing decline in profit rates, increasing institutional sclerosis, and an alarming expansion of the public sector. Recession has made resolution of these problems even more difficult and, with the exception of Japan, the adjustment mechanisms of major economies continue to react but slowly.

Higher oil prices have undoubtedly been among the principal

Table 2-3. Global Energy Demand (million barrels per day oil equivalent).

Energy Form	*1980*	*2000 (Estimate)*
Oil	61.6	70- 73
Gas	25.3	41- 48
Coal	34.8	55- 60
Nuclear	3.3	14- 17
Hydroelectric and others	9.9	17- 20
	134.9	197-218

causes of that recession, but while the direct effects of these are now beginning to recede, other factors have emerged to impede the normal recovery process, such as the sustained application of anti-inflationary policies, for example through expenditure cuts and tax increases to restrain budget deficits, except in the United States, and high interest rates designed to control growth of the money supply. Moreover, measures adopted by one country for understandable national reasons can have a depressant effect that can be rapidly transmitted to other countries via the world's closely linked financial system—a process accentuated by floating exchange rates and the importance of the dollar in international trade.

There are a number of natural corrective forces in any economy that could produce a normal recovery out of recession. Some such forces are the benefit to real wages from a fall in inflation, the investment stimulus that would come from lower interest rates if and when inflationary expectations subside, the need for replacement investment by industry, and the gradual rebuilding of stocks to normal levels. But the release of these natural forces depends on the ability of governments to reduce inflation and lower interest rates, and the way they approach these tasks and the degree of confidence with which investors view their efforts merely adds to the general feeling of uncertainty that pervades the economic outlook of the developed countries of the West at the beginning of the 1980s.

Taking these and other factors into account, in the event that an economic recovery does occur during the next five years, it seemed reasonable in early 1982 to envisage economic growth in the OECD countries averaging around 2.5 percent per annum through into the second half of the decade, with Japan and the United States doing rather better than Western Europe. The nonoil developing countries, which have shown more resilience than expected in adapting to balance-of-payments problems, occasioned partly by high prices and partly by reduced economic activity in the industrialized countries both in the West and the East, could average around 5 percent per annum growth over the same period, consistent with their trend in the 1970s. On the other hand, if economic recovery is delayed, OECD economic growth might not average much over 1 percent per annum with Western Europe again faring worse than either the United States or Japan, and although the non-OPEC developing countries would probably not be further greatly affected by developments elsewhere, their economic growth rates might be depressed to nearer 4 percent per annum or below. Much seems to depend upon the success or failure of the policies pursued by the U.S. administration, with the prospect, in the case of prolonged stagnation, of increasing unemployment, particularly in Western Europe.

While there are more than enough uncertainties surrounding the process of economic forecasting even over the next five years, when it comes to the longer term there are even fewer predetermined elements and therefore a greater element of unpredictability in the response of societies to the changes that got underway in national and global circumstances during the past ten years. While OECD economic activity has slumped in the wake of two oil price shocks, developing countries have fared comparatively well, although they have had to borrow heavily to overcome the adverse circumstances. This borrowing has, in many cases, simply been an alternative to a concerted effort to adjust their economies to the new environment. Furthermore, they are now suffering from high interest rates, debt servicing, and diminished export prospects as a result of the OECD recession. If the recession is prolonged, it is possible that the impetus for global growth that might result from a sufficient number of developing countries attaining the status of newly industrialized countries over the next two decades will not materialize. However, there are indications that trade between developing countries is on the increase, and if this trend is maintained, it could result in self-sustaining growth in the group of developing countries as a whole, possibly by the early to mid-1990s.

Energy is used mainly for cooking and lighting in agrarian societies, and noncommercial fuels such as firewood and vegetable wastes play a major role, with the result that annual consumption of commercial energy is as low as 0.5 to one barrel of oil equivalent per person. When nations enter the "take off" phase of development, per capita use of commercial energy rises sharply to five to ten barrels of oil equivalent per annum, as demand for energy-intensive products such as steel, cement, fertilizer, and chemicals increases. This compares with average per capita demand levels of some twenty to fifty barrels of oil equivalent per annum in fully industrialized countries. Reference has already been made to the special role that oil products have to play in the development process, and while nonconventional liquids, such as alcohol, may substitute increasingly for motor gasoline and diesel fuel in a number of countries, it is possible that more than half of primary oil demand in the noncommunist countries could come from the developing world as we enter the next century. It is true that substantial hydropower resources are being developed and that there is a large potential for biomass in the developing countries, but the demand for imported oil will nevertheless increase. This is yet another indication not only that it is imprudent to expect any single energy source to solve global energy supply problems but also that massive investment in all possible energy sources will be necessary if the envisaged volume increases are to materialize to meet the needs of both developing and developed countries.

Economic growth rates have in the past been important factors in evaluating energy requirements, but the relationship established between economic growth and increasing energy use is unlikely to provide much guidance for the next two decades now that the experience of the 1970s has stimulated energy conservation and substitution away from oil. We do not know how quickly improvements in energy efficiency, both in conversion and in end use, will take place; nor do we know when and to what extent consumers' urge to conserve and switch fuels will abate. There is already considerable uncertainty about the composition and level of final energy demand over the next twenty years due to structural economic change, technological innovation, changing lifestyles and consumer preferences, as well as the potential impact of government intervention via tax changes, subsidies, and so forth. The degree of improvement in end-use efficiencies in transport, industry, and domestic markets will be a critical determinant of OECD energy demand, and an important element in the end-use efficiencies will be the shift in the delivered energy mix toward electricity—from 16 percent of the 1980 total to an estimated 20 to 25 percent by the end of the century. There are certainly going to be sharp reductions in primary energy use per unit of GNP by the year 2000, but it is not clear at this time how great those reductions will be, and they will contrast with the higher levels of energy consumption per unit of economic activity expected in the developing world.

The uncertainties that attend any attempt to analyze the energy environment over the next two decades can, as already indicated, be adequately covered by three internally consistent scenarios that explore a range of alternative futures that could emerge. One scenario illustrates moderate economic growth and a fragile balance between oil prices and volumes; a second explores the highest feasible growth, with another oil crisis providing the catalyst for change, and the third describes a very low-growth world in which there is little restructuring of either economies or energy use and supply systems. The first scenario, which could represent "business as usual" and might last for twenty years, would require the absence of any serious discontinuity in oil supply or pricing. There would also need to be some restructuring in the economies of major consuming countries. Essential elements on which this scenario might be built would be, first, moderate growth in the global economy with Western Europe lagging behind the rest; second, a clear division in world trading patterns between Europe and the countries surrounding the Pacific Basin; third, significant improvements in energy efficiency; fourth, the continuing substitution for oil, with some active government support for alternatives; and fifth, the continuing cohesion of OPEC such as would ensure a stable oil market and an adequate supply of crude oil. Under this scenario, oil prices might steadily increase but in a manner that would not trigger radical responses on the part of either oil producers or consumers. For OPEC this would mean that "high absorbers" (i.e., those countries needing maximum-revenue earnings) would be able to fund their development plans, while the long-term markets for low absorbers would not be endangered. For the oil-consuming countries, a regime of orderly but moderate price rises would enable the existing momentum for conservation to be maintained and energy efficiency to be improved substantially. Continuing economic growth in this case would lead to increased oil demand, so that there would continue to be some incentive for further substitution away from oil and the development of alternatives, but there would probably also be rather limited government support for pioneering projects. In the absence of a further substantial shock to the system, Western Europe would continue to muddle along without tackling fundamental structural problems, but the economies of the United States and Japan would perform more successfully, as would those of the faster growing developing countries. As indicated previously, OECD economic growth could average 2.5 to 3 percent per annum with even Western Europe achieving almost 2 percent, and the developing countries would probably average 5 percent per annum. There could be substantial growth in both coal and natural gas supplies, and nonoil energy markets would develop fairly rapidly.

The second scenario envisages global restructuring in both economic and energy terms as a reaction to another oil crisis in the early 1980s. Essential elements in this case are, first, the highest feasible world economic growth; second, a rapid growth in the free international trade for most commodities; third, early and extensive efficiency improvements and strong moves to substitute away from oil; fourth, the rapid development of alternatives to oil within both national and regional energy systems; and fifth, considerable strains within OPEC that would make cohesion difficult. Under this scenario, energy would provide the necessary catalyst for economic restructuring. In the major OECD countries, recognition of the need for change would encourage innovation and risk taking. A strong sense of economic realism would prevail worldwide, and the process of adjustment would become one of self-sustaining growth as, for example, rising real incomes induced more tolerance for change and as resources were switched from consumption to savings and investment, from the public to the private sector, and from declining to growth industries. Greater flexibility in work practices and wage levels would increase both employment and profitability. World trade would be able to flourish in an open system and the axis of world trade would shift away from the Atlantic to the Pacific Basin. In such a vigorously competitive trading environment the developing countries would achieve rapid growth, with the emergence of many new newly industrialized countries providing a second engine for world growth. A rapid oil price escalation postulated for the early 1980s would provide a lasting stimulus for consumers to decouple from Middle East/OPEC oil, and there would be heavy investment in both conservation and oil substitution. OECD countries would emphasize less energy-intensive industries, and this would allow the high-intensity activities to move offshore, closer to energy sources. The combined effect of, on the one hand, the imposition by governments in the developed oil-consuming countries of stringent efficiency standards for new equipment and substantially increased taxes on oil products, and on the other hand, efforts in the developing countries to both limit the energy intensity of their economies and utilize potential indigenous supply sources would be to reduce pressure on oil supplies so that the market would once again be able to determine prices.

Under this scenario, the close interaction between output and worldwide trade would probably be sufficient to sustain good economic performance everywhere. Average GNP growth in the noncommunist areas could rise to nearer 4 percent per annum in the 1990s, and this would probably be the highest feasible growth in view of the predetermined constraints within the system in the early years. While this scenario depends largely upon international trade and specialization, the move toward national and regional self-sufficiency in energy implies that some markets for internationally traded energy could become constrained in the 1990s.

The third scenario is characterized by a protracted malfunctioning of most economies, and the resulting slump in oil demand would mean that, in the absence of any serious supply "accident," there would be no salutary price shock to the system such as would stimulate the restructuring referred to above. Essential elements of this scenario are, first, low economic growth and sluggish development of international trade, with the likelihood of widespread protectionism; second, only a slow improvement in energy efficiency and delayed substitution away from oil; third, little incentive to develop alternative energy sources; and fourth, a period of weak cohesion within OPEC, becoming stronger in the 1990s. Economies would continue to underperform, and nations would tend to focus upon domestic issues rather than the solution of global problems. Some of the OECD countries might succeed in muddling through, but they would have a chronic level of overcapacity and unemployed resources. International trade would be likely to stagnate, and large-scale protectionism would exacerbate North-South tensions. Western Europe would probably lead the protectionist movement, closing its doors to many imports from Japan as well as developing countries, and the latter would be seriously affected by the resulting decline in the volume of their exports.

For energy consumers, measures to improve efficiency already instituted would, under this scenario, continue to have a favorable influence, and governments would probably encourage conservation measures by increasing the tax element in oil prices in order to obtain much-needed revenue. But in the absence of further price signals, substitution away from oil would be likely to be relatively weak during the 1980s, and the development of alternative energies would be delayed. During the 1990s, the oil market could well begin to tighten due to rising demand in the energy-intensive newly industrialized countries. In such an environment, OECD economic growth could barely average 2 percent per annum, with slightly faster growth by the developing countries pushing the total average for noncommunist countries to some 2.5 percent per annum over the twenty year period. But a tightening oil market in the 1990s could well result in growing OPEC price aggression, and it could become very vulnerable to so-called accidents. A scenario of this type could mean requirements for very high oil volumes toward the second half of the period.

Although there were many elements of the third scenario recognizable at the time of writing, there was little firm indication as to which of the scenarios, if any, was more likely to influence the future of global energy supply and demand, and it should be born in mind that the implied change in energy consumption per unit of economic activity varies markedly between scenarios and from region to region. Within OECD, for example, energy intensities, particularly for oil, show large decreases over the period due to continued improvements in conservation and end-use efficiency, a shift in economic activity to less energy-intensive goods and services, and substitution for oil. But, under all scenarios, energy intensities rise in the developing countries, and taking the two together, primary energy demand in the noncommunist world seems likely to rise from 94 million barrels per day of oil equivalent in 1980 to a range of 135 to 150 million barrels per day of oil equivalent by the year 2000. The difference in the scenario range will depend mainly on the extent of government intervention and upon improvements in end-use efficiencies, which would be highest in both cases under the second scenario. In the OECD countries the changes in the fuel mix across the major markets are likely to incorporate the substitution of gas and fuel oil by coal and nuclear energy in the rapidly growing power generation sector, the increasing penetration by coal and natural gas of the industrial market, and in Europe and the United States, the domination of the domestic sector by electricity and natural gas. A decline in primary oil demand in the three major consuming areas of the OECD will most affect demand for heavier products as, in power generation and under-boiler outlets (i.e., steam generation), it is substituted by increases in coal and nuclear energy.

The time has now come to look more closely at the place of natural gas in the global energy picture, to see how it has evolved as a significant contributor to the fuels/energy balance of many countries, to examine its resource base, and to consider the extent to which it may continue to be mobilized in spite of all the uncertainties attending the unfolding of events during the last two decades of the twentieth century.

3 NATURAL GAS IN THE WORLD ENERGY PICTURE

So much knowledge is available on the origins and development of the international oil industry that many people when thinking about international energy flows tend to concentrate upon oil to the virtual exclusion of natural gas. Perhaps a very brief historical look at the origins of the natural gas business will help to counteract this myopia. Since an excellent account of the history of both the manufactured and natural gas industries already exists,[1] I will herein give only a brief account of how natural gas has come to play the part in the global energy supply that it plays today.

Whereas the ability to make a combustible gas by heating coal in a retort was discovered toward the end of the seventeenth century, coal gas was not apparently used by anyone for effective domestic purposes until near the end of the eighteenth century. The first commercial sale of coal gas did not occur until the beginning of the nineteenth century, initially in the United Kingdom but soon thereafter in the United States and the German States. By the middle of the nineteenth century, gas manufactured from coal was lighting the cities of Brussels, Sydney, and Toronto, and by 1875 it was being used on a significant commercial scale in Japan, India, Mexico, Argentina, Poland, Italy, Denmark, Spain, and other countries.

The first recorded commercial use of natural gas was in New York

1. Malcolm W. H. Peebles, *Evolution of the Gas Industry* (London: Macmillan, 1980).

in 1821, and by 1861 it was also being used slightly in Canada. Although manufactured coal gas and natural gas were to commence commercial development in the United States at just about the same time, coal gas was for many years used primarily for illumination, while natural gas was used mainly for heating and industrial purposes. Sales of manufactured coal gas continued to expand in the United States until the middle of the twentieth century, although today almost all of the gas used is natural gas. The first recorded large-scale industrial use of natural gas was in Pittsburgh in 1883, where it was also used for domestic purposes a few years later. Initially, natural gas in the United States was confined to outlets close to the producing fields since transportation and distribution technology was no more highly developed than were production techniques, and much of the gas came, therefore, from small shallow fields, the reserves of which were often rapidly exhausted. During the fifty years following the Pittsburgh development, growth of the natural gas industry was slow, and use was limited to those states where it was produced, primarily in association with crude oil. Elsewhere, particularly in oil fields where there were few if any local marketing prospects, enormous volumes of this associated gas were flared, and the volumes thus wasted in the United States continued to grow until the 1950s. It was with the advent of the availability of high-tensile steel pipe in large quantities that the long-distance transmission of natural gas became possible, by the mid-1920s, and by the end of World War II huge natural-gas deposits in Texas had access to the major markets on the eastern coast of the United States. By 1966, natural gas was available throughout all states south of the U.S.-Canadian border, and manufactured coal gas was sharply on the decline.

At around the turn of the century, electricity had been introduced for lighting, and gas lost its domestic role in the United States, but it, in turn, quickly displaced coal in the cooking and water-heating markets and, thus, presented a good illustration of how the advancement of one form of energy in a market or end-use sector is almost always at the expense of some other form of energy. It is interesting that after World War II, outside the United States, manufactured coal gas was displaced in many end uses not directly by natural gas but by oil, and indeed natural gas only contributed barely 2 percent to total primary energy consumption in Western Europe in 1965. It has only been during the last two decades that natural gas has achieved rapid growth in the West European markets where, today, it meets something over 14 percent of energy demand. Oil was the major growth fuel after the war, and the manufactured gas and coal industries saw their markets being eroded in both industrial and domestic uses. Manufactured gas was becoming regarded as an old-fashioned form of fuel, and in many countries sales and volumes started to decline even though, by the late 1950s, gas-making processes based on oil feedstocks began to be commercially adopted. Gas manufactured from coal is a relatively low calorific value product and, when natural gas became widely available in the 1950s and 1960s, gas companies not only had to adapt consumers' appliances to accept the higher calorific value gas with different burning characteristics, but they also had to change their approach to the handling, distribution, and marketing of this new product; a new infrastructure had to be created virtually overnight. It is to the considerable credit of the gas industry that natural gas was introduced so successfully in so many countries in a comparatively short time and in a manner that resulted in a minimum degree of disruption to gas consumers in particular or to the public in general.

From the point of view of the wastage of a finite source of hydrocarbons, it was most unfortunate that the true value of natural gas was not appreciated during the early days of exploration for oil. Oil was what the drillers were looking for, and if they found either gas unassociated with any oil deposit or overlying or associated with oil, it was considered something of a misfortune. In any case, the markets did not exist nearby or on a large enough scale to make the collection and distribution of the gas commercially attractive, given the technology and equipment that was available in the late nineteenth century. It is horrifying, in the 1980s world of tightened hydrocarbon availability, to contemplate how much natural gas must have been burned or vented into the atmosphere during the last one hundred years but, even with today's vastly improved technology and understanding, a great deal of this premium fuel is still escaping from man forever, for mainly financial reasons. But how much natural gas is there left, despite the wastage and the rapid build-up of consumption over the past few decades? And how much longer can it last if more and more people and countries use it?

In a major study on the global situation, the International Institute for Applied Systems Analysis (IIASA) has suggested that around 8 percent of total estimated natural gas resources have been consumed to date, compared to 15 percent or more for oil.[2] But whereas total global oil resources, considered ultimately available, are thought to have been fairly accurately estimated by now, total gas reserves are continuously being reevaluated upward and many new discoveries of unassociated gas deposits are still being made. It is widely believed that a lot more gas will be found when people look for it specifically instead of finding it as a by-product in the search for oil. Furthermore, as technological advances enable deeper drilling to be carried out, so the likelihood of discovering more natural gas will increase. Meanwhile, the trend of finding new giant or supergiant gas fields is not reversing. The recovery of natural gas from proven deposits is high, 80 to 90 percent in many cases, a substantially higher rate than for oil. IIASA believes that as much as 280 trillion m^3 remain to be produced, of ultimately recoverable conventional gas alone, and that gas production globally could increase to between 3,700 and 4,000 mrd m^3 during the first decade of the twenty-first century—a level that could be maintained for several decades. Other sources believe that future discoveries of exploitable conventional natural gas could be anywhere between 60 trillion m^3 and 220 trillion m^3, but they point out that the most prospective regions are the Middle East and the Communist areas which, between them, might provide some 70 percent of the new discoveries, which are likely to be mainly of nonassociated gas.

The current level of natural gas production is around 1,600 mrd m^3, of which nearer 1,450 mrd m^3 is actually consumed. World proven reserves of natural gas are probably by now in excess of 80 trillion cubic meters, and there is thus a reserves-to-production ratio in the region of 52 to 1. By the year 2000, remaining recoverable global gas resources could still be at least 160 trillion m^3 and this should eliminate any doubts about the important role that natural gas can play in the transition period from fossils fuels to renewable energy, provided always that investment and development decisions are taken in time and that there are no political restrictions that seriously hinder the realization of the numerous projects that will have to be involved.

The transition to renewable energy will take place over a period in excess of fifty years, during which the world will have to concentrate on lower grade fossil fuels than those on which it has become

2. IIASA, *Energy in a Finite World*, Vol. I, *Paths to a Sustainable Future*, and Vol. II, *A Global Systems Analysis* (Cambridge, Mass.: Ballinger Publishing Company, 1981).

accustomed to rely over recent decades. It is difficult to see how this can be accomplished without much greater pressure on such a convenient fuel as natural gas. Natural gas is one of the few forms of primary energy that can be used as final energy, and that is why it is a fuel of preference when it is available. It is the international availability of natural gas over the next few decades that is therefore one area to concentrate upon, for the global demand for gas ought undoubtedly to increase as communities strive not only to solve their immediate energy problems but also to set in place strategies for the future. The production of natural gas obviously could grow faster, to keep pace with increasing demand, but whether or not it will do so depends upon numerous things, some of which will be considered in more detail in respect of Western Europe.

The IIASA study suggests that since reserves of both conventional and unconventional—that is, ultra-deep or ultra-low-grade gas or gas from hydrates or tight reservoirs—gas are so large that natural gas can take an ever-increasing share of the global energy market, maybe even reaching a peak of over 50 percent during the first decades of the twenty-first century, particularly if the cost of long-distance transportation can be reduced. It is, in reality, most unlikely to do this, however, in view of competition from electricity, generated increasingly from coal, and by nuclear and solar energy. However, if political and environmentalist opposition brings about a moratorium on the development of nuclear generated electricity, gas would be ready and able to increase its share, which might otherwise be limited, especially in the field of secondary energy demands.

Because natural gas is both market and transport limited, it is unlikely that it will in fact contribute much more than 20 percent of global energy supply by the year 2000—a share similar to its current contribution. It should be stressed that, in volume terms, this alone will require that current production levels be more than doubled. For reasons that will hopefully become clear, the future development of global gas consumption is rather more likely to be market limited than supply constrained although there is a more intricate relationship between supply and demand for natural gas than there is for oil. A 4 percent increase in global annual consumption, such as was achieved in 1980 over 1979, should therefore probably not be expected to be repeated very often for the foreseeable future. Indeed, for a number of reasons, natural gas, in spite of its comparative global abundance, may never have an era of its own as did coal and oil, and even though consumption will increase significantly in some countries, it will stagnate or even decline in others. This will particularly apply to those countries in which is has already achieved a high level of market penetration, as in some Western European countries, and in the United States, in view of conservation forces and government measures aimed at energy supply-source diversification.

Since this book is primarily concerned with the long-term supply of natural gas to Western Europe, interest must focus on domestic European production potential and upon those countries currently able or likely to be in a position to export to Western Europe. However, some comments on the general global supply and demand situation from time to time would not be out of place. In 1980, with world proven reserves estimated at well over 75 trillion m^3, the Soviet Union possessed over one-third of these and the Middle East, notably Iran, Qatar, and Saudi Arabia, another third. Other major reserves-holding countries include the United States, Algeria, Canada, Mexico, the Netherlands, and Norway, and there are substantial reserves in Australia, Indonesia, Malaysia, and West Africa. It is quite possible that China also has large resources, but details of these are not well known. However, merely possessing large reserves by no means implies that a country either can or will be a gas exporter. Some of the larger reserve holders, such as the United States, the Soviet Union, Canada, and the Netherlands, already have well-developed domestic markets for natural gas and either have no possibility to export (e.g., the United States) or are exporters but face dwindling export prospects when reserves pass peak productive capacity (the Netherlands). The Soviet Union and Canada, however, can increase exports even though they have growing domestic markets. Other countries with very large reserves, such as Saudi Arabia, Nigeria, Iran, and Norway, have small or totally undeveloped domestic markets and are clearly likely to be able to export gas for a long time to come. Whether or not they do so depends upon political as well as strictly economic decisions. If a country has significant proven reserves—that is, reserves numerically sufficient to allow for an export surplus—it may not actually be in its interest ever to export any in view of, for example, its distance from the nearest worthwhile market, its own determination to reduce imported energy by developing domestic sources, or the inconvenience of the location of its reserves such as would make them too costly to develop for export. So it is not particularly useful to list the major reserve holders, along with their current production levels, and suggest that such and such a reserve-to-production ratio would indicate an export certainty. Indeed, the "rigidity" so often ascribed to gas and resulting largely from transportation constraints suggests that the value of a particular gas find depends very much upon just where, locally and globally it is located. In this connection, coal and oil are far more flexible and, when comparing reserves figures, the ones for natural gas have to be discounted.

An additional factor when assessing the export prospects for a natural gas reserve-holding country is that its gas may be one or a combination of four different types: gas that is associated with oil (either "wet" gas, which contains substantial quantities of gas liquids, or "dry" gas, which contains no gas liquids), or unassociated gas (again either wet or dry). Also, much will depend upon whether or not it is already an oil-exporting country. If, as is the case with Angola, it already exports oil and its gas is mainly associated gas, it can well be the case that it will not wish to commit itself to a gas export scheme because this could limit its oil export volumes and even pricing. Similarly, since associated gas is produced in accordance with the volume of oil produced, if too much oil is produced, gas would have to be flared if the gas export scheme could not accommodate it, as has been the case with Nigeria. On the other hand, too little gas might be available to meet export commitments if oil production were to be cut below appropriate levels, for market or for political reasons. Such a country might be better off, depending upon its economic circumstances, developing its associated gas for domestic purposes, thereby extending the life of its oil export potential by reducing domestic demand for oil. A country such as Algeria, with large unassociated natural gas reserves, that is also an oil exporter may, again, wish to develop a domestic market for its gas. But if it is faced with declining oil exports over the years, it could make sense to develop an export market for its gas as well. A country with a lot of unassociated gas but little or no oil production capability, and which therefore has to import, could be expected to develop ways of using domestically as much gas as possible in order to minimize foreign currency expenditure upon oil, and only then develop export possibilities.

It can be seen from all this that it is unwise to generalize about the export potential of gas-producing countries, and every such country must be assessed individually. With this in mind, there are a number of countries that, because of existing proven reserves, are

Table 3-1. Breakdown of International Trade in Natural Gas (percentage shares).

Main Exporters	*1980* *100*	*1990 (Estimate)* *100*
USSR	26	25
Netherlands	24	7
Norway	14	7
OPEC[a]	14	32
Others[b]	22	29
Total volume (approx)	*190* mrd m^3	*420* mrd m^3

a. Including Algeria, Abu Dhabi, Indonesia, Iran, Libya, Nigeria and Qatar.
b. Including Canada and Mexico.

certainly able to increase their domestic use considerably, and some of those may, in addition, be likely starters for gas export schemes. Angola, India, and Cameroon, for example, have mainly associated gas. Tunisia, Egypt, Syria, and Brunei have both associated and nonassociated, and Trinidad, Columbia, Ecuador, Argentina, Bahrain, Oman, Pakistan, Qatar, Bangladesh, Thailand, and New Zealand have mainly nonassociated.

The countries that are currently exporting natural gas are numerous, the most significant being the Soviet Union, the Netherlands, Norway, Canada, Indonesia, Mexico, Algeria, Brunei, Afghanistan, Iran, and Libya. Internationally traded gas in 1980 amounted to approximately 190 mrd m^3, of which half was supplied by the Soviet Union and the Netherlands, and only 14 percent was supplied by countries of OPEC, comprising Algeria, Abu Dhabi, Indonesia, Iran, and Libya. Nigeria and Qatar are expected to join the OPEC exporters within ten years, and OPEC might then account for almost one-third of internationally traded supplies, which may then total as much as 400 mrd m^3. The Soviet share will still be in the region of one-quarter of supplies, but that of the Netherlands will have fallen to a mere 7 percent, while Algeria and Indonesia will provide 8 and 6 percent, respectively. Of the 1980 total of 190 mrd m^3, 83 percent was delivered wholly via pipeline; only 17 percent in the form of liquefied natural gas (LNG), required transport by tanker for most of its journey from wellhead to burner tip. By 1990, LNG's share is expected to have risen to nearer 40 percent, but much will depend upon whether sufficient project investment decisions can be taken in

Table 3-2. Percentage Breakdown of International Trade in Natural Gas between Pipeline and LNG Deliveries.

	1980 *100*	*1990 (Estimate)* *100*
Pipeline	83	61
—of which OPEC supplied	2	8
LNG	17	39
—of which OPEC supplied	12	24

the very near future, at a time of considerable international political and economic uncertainty. (See Tables 3-1 and 3-2.)

Having briefly reviewed the producing and exporting countries, it is now time to turn to those countries largely responsible for consuming all the gas and for providing markets that, depending upon one's viewpoint, can be seen as either convenient outlets for energy resources that might otherwise not be exploited or as hapless victims of the producer countries. The established consumer countries are already locked into a form of energy dependency by virtue of their costly and extensive gas industry infrastructures, which cannot be allowed to remain underutilized, let alone completely idle. Furthermore, a large part of the consumption is effected by domestic consumers, otherwise known as potential voters. Some 1,450 mrd m^3 was consumed globally in 1980, of which 63 percent was by the United States and the Soviet Union. Western Europe consumed another 14 percent, Canada 3.6 percent, and Japan 1.8 percent. While consumption by the United States is likely to fall somewhat from the 1980 level of over 540 mrd m^3, that by the Soviet Union will rise from 380 mrd m^3 to perhaps over 600 mrd m^3 by 1990, when that country will probably be both the largest producer and the largest consumer. Depending upon which scenario environment for economic growth and restructuring prevails, Western Europe might be needing up to 270 mrd m^3 by the 1990s, compared to just over 200 mrd taken today, and Japan's imports should have risen well above the 1980 level of 26 mrd m^3. Global consumption could have risen to around 1900 mrd m^3, and a substantial increase is expected to occur in developing countries, especially among the natural-gas producers themselves. Among the countries of the Organization for Economic Cooperation and Development (OECD), rising demand is expected in Australia, New Zealand, Canada, France, Italy, Spain, Denmark, Greece, and Sweden, and both South Korea and Taiwan could follow Japan's example and become significant LNG importers.

In the longer term, the international gas trade is likely to continue to develop on a regional basis in view of logistical and other factors that will be described later. By the year 2000, for example, 65 to 70 percent of northwest Pacific Basin LNG imports are expected to be supplied from Southeast Asia and Australia, whereas 70 to 80 percent of European imports could come from the Soviet Union and Africa. Gas from the large Middle Eastern deposits could go either to the Far East or to Western Europe, but most likely to the former for reasons that will also be dealt with in more detail later. Growth in Japanese imports of LNG centers initially on the future growth of its use as a premium fuel for urban power plants, with the industrial and public distribution markets becoming more important in the longer term. New markets in South Korea and Taiwan may eventually be largely premium residential and industrial, and the bulk of supplies will originate from the Southeast Asian and Pacific Basin countries, of which Indonesia will probably be the largest supplier, supplemented by supplies from the Middle East. Demand in the North American continent will largely be met by supplies from countries in the Western Hemisphere, with Mexico and Canada increasing their exports to the United States as well as, possibly, Japan and even Western Europe. The United States could one day possibly take increased LNG supplies, which are currently small, from North and West Africa, but there are a number of political uncertainties attached to this option. If the United States ever does become a significant LNG importer, this could have repercussions upon the West European position, but at present the indicators are that the West European and Western hemisphere gas markets will remain largely free of each other's influence. Central and South America may well become significant and growing users of natural gas, but it seems likely that their requirements will also be met on a regional basis even though Brazil might import from Africa. The Middle East and West African markets are for the foreseeable future unlikely to face supply problems that would depend for solutions upon countries outside their region. There can of course be exceptions to all these regional trends and, for example, it might be possible to import natural gas produced in Argentina or Trinidad or, more likely,

Canada into Western Europe, but this currently seems a less likely development given the uncertainties facing the development of international trade generally and the energy industries in particular. Much will depend upon pricing considerations and the needs of individual producer countries to earn foreign currency.

While international trade in natural gas is now usually based on prices that are indexed in a variety of ways to competing oil products, this is far from being the case with gas prices in general, and it should be emphasized in this connection that over 85 percent of world gas consumption still consists of indigenous gas, which is priced at widely different levels, often for historical reasons but also depending on local circumstances and government policies.

These aspects will be considered in the next chapter. For now, some general comments on the three main consumption areas outside the Soviet Union and Eastern Europe will serve to indicate the sort of thing that is involved. In North America, the dominant issue will clearly be the timing and the extent of the decontrol of indigenous U.S. gas prices, which are, on average, well below market levels. Natural gas in 1981 constituted 25 percent of the total energy supply of the United States, but by the end of the decade many people expect that share to have fallen nearer to 20 percent. In 1982, over 93 percent of the total 540 mrd m^3 demand for natural gas, some 510 mrd m^3, was produced in the continental United States, while 6 percent was imported from Canada and Mexico by pipeline and 1 percent from Algeria as LNG. Nearly 40 percent of U.S. demand at this time was for industrial use, 40 percent for residential and commercial, with the remaining 20 percent used as chemical feedstock or by electric utilities. An increase in residential and commercial uses over the next ten years is forecast to accompany a small decline in industrial and electricity-generation uses, so that overall gas demand is expected at most to remain at current levels or to decline somewhat.

One of the factors contributing to the envisaged decline in U.S. supply and demand for natural gas in the longer term is the declining reserves-to-production ratio, which is now around eight years, having been over ten years in 1972. Additions to reserves still continued to lag behind production in 1982, only reaching about 400 mrd m^3 or 70 percent of demand annually, and it was clear that current conventional production levels of over 540 mrd m^3 per annum could not be sustained for very long. It was unfortunate, from the point of view of encouraging new exploration efforts, that there appeared in 1982 to be something of a natural gas glut, or temporary excess deliverability, but this was expected to phase out over the next few years owing to changing field development patterns in the United States, which would require fast depletion of some new fields, for example those in the Gulf of Mexico. By 1990, United States production could be less than 450 mrd m^3 and, unless major new conventional resources are discovered, the decline can continue to the end of the century, when less than 400 mrd m^3 could be being produced. If demand declines even below 500 mrd m^3 it seems likely that there will still be a substantial shortfall of conventional indigenous natural gas supplies developing over the next two decades and that imports will have to increase substantially over the current level of 42 mrd m^3 per annum.

The significance of the decontrol issue in the United States is that the sooner the domestic price of natural gas rises to market levels, the greater will be the pressures to conserve gas and the greater will be the incentives to find and produce more domestic gas, since the wellhead prices will become higher. The Natural Gas Policy Act of 1978 (NGPA) was enacted to overcome what was, even at that time, perceived to be an incipient gas supply shortfall. Reduced demand was to be achieved by restricting the use of natural gas for under-boiler fuel purposes through the Fuel Use Act, and the encouragement of supply was to be achieved by gradually lifting wellhead price restrictions so that, by 1985, all wellhead prices except for "old gas"—that discovered prior to 1977—would be deregulated. The escalation of natural gas prices that was to bring new gas prices into line with the market by 1985 appeared, by 1980, to be completely inadequate, following the second "oil shock" of 1979, since it was based on the assumption that the price of oil would be around $15 per barrel in 1985. In early 1982 it appeared that, instead of the planned smooth transition in 1985, gas prices would probably jump significantly on deregulation, even with world oil prices then falling from the OPEC market price of $34 per barrel. Some rephasing of gas price escalation was generally thought necessary, but it had not yet been agreed how this might best be achieved. For example, price increases could be speeded up before 1985, or the year of full deregulation could be extended beyond 1985. As for the deregulation of old gas—which producers would like to see on the grounds both that it would promote the better use of a scarce resource and that the additional cash flow generated would provide for additional efforts to discover and produce more—there seemed, by mid-1982, no prospect of a major legislative change before the end of the year. Old gas was still expected to constitute 40 to 50 percent of supplies in 1985. Opponents of the deregulation of old gas maintained that it would result in enormous windfall profits for the producers, and they proposed that any such deregulation be accompanied by a windfall profits tax, but, at the time of writing, it was clear that this politically sensitive issue was to be shelved until after the 1982 congressional elections, particularly following the long, hard winter of 1981/82 which resulted in sharply increased home-heating bills. Even if old gas is not deregulated in due course, there will probably be some reduction in overall consumption. A sharp rise in gas prices, generally, would accelerate this to the point where prices might even have to be reduced somewhat to prevent further erosion of gas markets. All this could have relevance to the international gas trade in that it can influence the extent to which the United States may have to import supplemental gas in the form of LNG from certain producer countries that would otherwise be limited to finding export markets in Western Europe. As things stood in respect of deregulation in 1982, it appeared that there would be a depressing effect upon the U.S. gas market share, at least in the short term, which would tend to reduce the attractiveness of imports, particularly of LNG.

If West European gas prices were to stabilize at their relativity to those oil prices obtaining in 1982, the prospect is that there will be a slow but steadily growing demand for imports, from whatever source; hopefully from as many sources as possible. If higher prices should materialize, a substantial move away from nonpremium markets could occur, with a subsequent weakening of demand and, ultimately, a possible downward readjustment in prices. There would, in effect, be a transition in the European gas industry from a cheap and readily available fuel to a more highly valued energy source.

In 1982, Japan appeared to be facing a temporary reduction in demand growth rates, with gas demand for power generation moderating as a result of conservation measures, at a time when new supplies, from Malaysia, were due to develop in the short to medium term. However, if gas companies succeed in penetrating residential and industrial sectors more deeply, this need not be the case. The

North Pacific Basin area, where Japan is by far the largest energy user, is dependent almost entirely upon imports and, unlike the United States and, to a lesser degree, Western Europe, has perforce become something of a seller's market. If new markets were to develop in South Korea and Taiwan, it would tend to perpetuate this condition, although if natural gas is ever to increase its penetration in nonpremium market sectors in the area, some weakening of gas prices relative to oil may be necessary. Unlike the major consumers mentioned above, Japan is the only major industrialized area of the world with very small reserves of indigenous energy. Local production of natural gas remains negligible, and the market development has been based almost entirely on the developments of imported LNG. Although overall natural gas consumption remains small in world terms (less than 30 mrd m^3 or 2 percent of world consumption in 1982) there are already some 15 million customers, and it is thought likely that Japan will wish to acquire increasing quantities of LNG in coming years. Japan has in fact had a traditional manufactured gas industry for a hundred years, which, prior to the introduction of LNG, had moved from dependency upon coal to newer forms of feedstock such as naphtha and LPG and which supplied large numbers of customers at low levels of household consumption. The major change began with the conclusion of a contract in 1966 for the supply of LNG from Alaska to Tokyo Electric Ltd. and Tokyo Gas Ltd. It is interesting to note from the outset that the majority—75 percent—of the delivered quantities were purchased directly by Tokyo Electric, the supplier of electric power to the Tokyo area. Uniquely, because of the pollution problems surrounding major cities, severe regulations had been imposed which, in effect, forced major fuel users to move to so-called "zero sulphur fuels" such as low-sulphur crude, LPG, naphtha, and so forth. In comparison with these, LNG offered an attractive alternative. Following this small initial supply, larger quantities have been contracted from Brunei, Abu Dhabi, and Indonesia, which was in 1982 the largest exporter of LNG in the world. Customers were still, typically, groupings of power companies and gas utilities. Among the latter, the three largest—Tokyo Gas, Osaka Gas, and Toho Gas—resolved to convert all their customers from manufactured to natural gas. Work began by Tokyo Gas in 1972, and the others followed suit shortly thereafter. Conversion is expected to be completed by the mid-1980s. Outside the electricity-generation sector, however, the penetration of gas remains relatively small. This is due to a combination of Japan's mountainous and difficult terrain, which allows for only limited pipeline building; price controls by government on kerosene, which is the main competing fuel in the residential sector; and gas prices that have proved too high for the industrial sector. It is likely that attempts will be made to remedy all of these during the coming years, since Japan is keen to expand its use of LNG in order to reduce its degree of reliance on imported oil. As a result, Japan is likely to retain its position as the world's leading importer of LNG.

One major energy-consuming area that remains to be dealt with is Eastern Europe. It is not particularly significant in terms of internationally traded natural gas insofar as it buys all its gas imports from the Soviet Union, but it does lie directly across the pipeline route for Soviet gas exports for Western Europe, and political and economic developments there can have a considerable influence upon the timing and extent of East-to-West energy flows. With the exception of Romania, which once had considerable hydrocarbon reserves–sufficient to sustain a domestic production of as much as 14 million tons per annum of crude oil and over 30 mrd m^3 of natural gas annually–all the other East European economies were essentially coal based. Only during the last three decades, with the growing availability of Soviet hydrocarbon exports, did the necessary investment take place to facilitate the greater use of these more efficient energy forms.[3] Although several of the countries–Poland, Hungary, and East Germany in particular–do produce some natural gas, there has been insufficient supply to meet their growing demands. Thus, they imported, along with Yugoslavia, some 35 mrd m^3 of Soviet natural gas in 1980 and around 80 million tons of Soviet crude oil. This still left them with the need to import some 15 to 20 million tons of oil from elsewhere, but, in view of the virtual isolation between 1940 and 1970 of their economies and financial systems from those of the rest of the world, they found the burden on their scarce resources of hard currency extremely heavy. All of these countries except Czechoslovakia have borrowed heavily from the West in order to finance their imports not only of energy but also of a wider range of industrial and consumer goods and technology. By the end of 1981, their combined net indebted-

3. Jeremy Russell, "Energy Problems in the Soviet Union: Problems for COMECON?" *World Economy* 4, no. 3 (September 1981): pp. 291–313.

ness to Western private or government financial circles was in excess of $60 billion. Their reliance upon Soviet energy supplies, which prior to that time were purchased almost entirely through other than hard-currency payment, is therefore extremely important for their economies.

In recent years, however, the Soviet Union has let them know that the mainstay of their energy imports, Soviet crude oil, is not to be available in the foreseeable future in quantities larger than those supplied in 1980 and that even some of this may have to be paid for at world prices instead of at the heavily subsidised rates of the past, which, in some years approached 50 percent below world prices. The reason for this oil shock to the East European system is that, not only has the yearly rate of increase in Soviet oil production declined to around 1 percent while Soviet consumption continues to increase by some 2 to 2.5 percent, resulting in a dwindling surplus of exportable oil, but the Soviet Union itself also has a large and enduring requirement to earn hard currency and has allowed hydrocarbon exports to grow to the point where they now account for over 60 percent of all Soviet hard currency earnings. The Soviet Union is therefore faced with the chronic dilemma of dividing its oil exports between its allies paying soft currency and its hard currency markets in the rest of the world, mainly in Western Europe. During the last two decades the East Europeans have been taken increasing quantities of Soviet natural gas, and in view of the now almost inevitable stagnation or even decline in their access to Soviet oil, natural gas will play an increasingly important role in their energy imports.

Soviet gas is delivered via a network of pipelines from fields in the Ukraine, and from further to the east, in Orenburg, western Siberia, and the Komi Republic. In many cases the construction of parts of the main pipelines has been achieved with material and personnel assistance from the East Europeans, who have been obliged to make considerable investments in the development of the Soviet gas and other energy industries in order to ensure their future energy supplies. Much of the gas now flowing from the Soviet Union to Eastern Europe does so along routes that also carry natural gas on its way to the hard currency export markets in Western Europe. Although it might only be a remote possibility, even taking the Polish situation into account, any major unrest in the area that required military action on the part of the Soviet Union to stabilize could bring about a disruption of supplies to those West European markets.

The East-to-West flow of Soviet oil and gas to both Eastern and Western Europe, its significance for both the East European economies and balances of payments and for Soviet hard currency earnings, and its vulnerability to interruptions arising from political or economic unrest in the East European countries is something that perhaps also needs to be taken into account when consideration is given to other aspects of West European energy import dependence upon the Soviet Union.

This chapter has looked at some of the broader aspects of natural gas in the world energy supply picture as well as at some of the more specific cases, and it has attempted to look at the part that natural gas may play in the energy scenarios of the future. It is difficult to escape the conclusion that, at the moment, natural gas is not contributing as much to meeting global energy demands as it could, and that this is largely due to the difficulty of transporting it from where it is found to the countries that, for whatever historical reason, have well-developed markets for it. Its underactive contribution is, however, partly due to the dominating role that oil has played in the world energy scene for many decades and to the fact that there are certain end-use markets where natural gas is not yet able to play a significant role, such as the automobile and other transportation sectors as well as those where it has to compete with readily available alternative energy forms, such as electricity based on coal and nuclear energy. A glimpse has been taken at some of the national and international commercial and political considerations that have to be taken into account when examining the future of internationally traded natural gas and at some of the interrelationships influencing the trend toward regionalization in the development of that trade. But the time has come to look more closely at natural gas itself, at the key factors of the gas industry, the markets, the infrastructural requirements, pricing, and the problems of transportation and distribution, all of which play an important part in the matching of supply with demand.

4 KEY FACTORS OF THE NATURAL GAS INDUSTRY

If it is accepted that, in terms of energy equivalence, global reserves of natural gas are of a similar order of magnitude as those of oil, it is fair to ask why it is that this gaseous hydrocarbon has not made a similar contribution to world energy supplies and why it does not at present seem likely that it ever will. In this chapter we will look at some of the main differences between oil and natural gas that affect the extent to which the latter is likely to achieve penetration of the world's energy markets and at the factors favoring or limiting its potential to participate in international energy trade over the next two decades.

It is appropriate to start by recognizing the limitations resulting from the basic molecular structure of the chief component of natural gas: methane. Methane (CH_4) is relatively unreactive chemically, and this limits its use in many of the major petrochemicals markets, but CH_4 is also a very simple hydrocarbon molecule that, although it is extremely good for combustion, is limited in the range of nonfuel applications to which it can be put—applications for which oil, with its much more complex composition, has found ready acceptance. Bitumen, liquid fuels, and lubricants for transportation are a few such applications. On the other hand, natural gas burns much cleaner than do fuels derived from coal or oil, which contain traces of sulphur. Thus, the gas that is finally delivered to customers causes the minimum of pollution on combustion in domestic or industrial appliances. This makes natural gas in demand not just for those applications where clean air and environmental safeguards are a major consideration, but also in a wide range of industrial processes where it is necessary to avoid contamination. Where natural gas is not excluded by its chemical and physical properties, therefore, it is a very attractive fuel providing an easily controlled rate of heating, an odorless combustion, and, where required, high heat intensity. Moreover, because it is piped direct to the point of application, it needs no customer storage or other handling equipment so that, over and above its inherent thermal efficiency, it finds an extensive range of potential markets. Indeed, it can compete in all the fuel markets, which typically make up around 75 percent of the primary energy consumption in an advanced country in the Northern Hemisphere. It can, in particular circumstances, even be used in transport markets, either converted into methanol or, as a result of recent technological advances, gasoline. It can also be used directly in compressed or liquified form.

In its gaseous form, natural gas is clearly a far less dense energy carrier than oil and it is therefore substantially more costly to transport to its markets. The "cost of distance" is one of the major limitations on the development of market penetration by internationally traded natural gas and therefore merits some elaboration. For oil, increasing distance is comparatively unimportant when it can be transported by large-diameter pipeline or by a very large crude carrier, but for gas the position is rather different. In most cases a pipeline is the logical and most practical method of transporting natural gas between the wellhead and the market, but comparative analysis shows that even with the use of a 56-inch (1.4 meters) pipeline, which approaches the maximum currently in use in international trade, the cost of transport rises much more rapidly with distance than it does for oil. A major reason is the need for additional compressor stations to boost the flow of gas over a long distance. If the pipeline is to be laid in difficult conditions—for example, underwater—the transmission cost increases still more rapidly because of the higher capital costs. This has proved one of the major deterrents to the international trade of gas because a substantial proportion of the world's proven natural gas reserves are separated from potential customers either by many thousands of miles or by oceans beneath which it would be infeasible, even with modern technology, to lay

Figure 4-1. Comparison of Oil and Gas Transportation Tariffs.

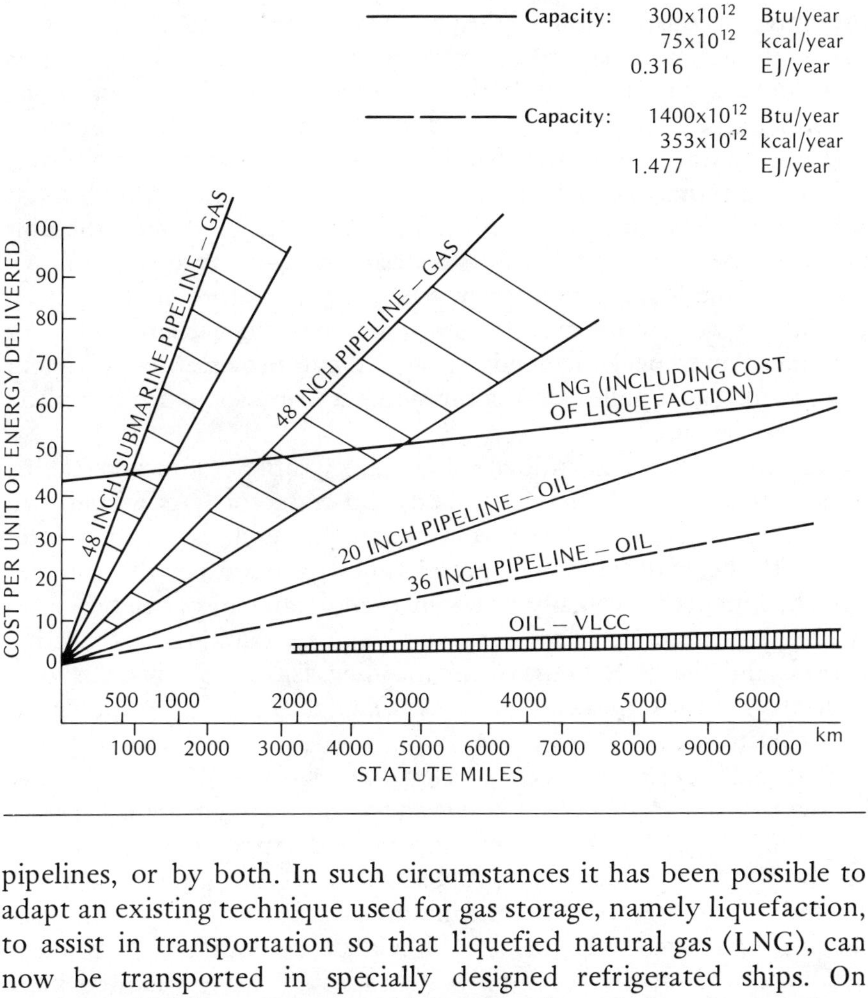

pipelines, or by both. In such circumstances it has been possible to adapt an existing technique used for gas storage, namely liquefaction, to assist in transportation so that liquefied natural gas (LNG), can now be transported in specially designed refrigerated ships. On account of high capital cost this method is uneconomic over short distances, but because the overall transport costs rises much less rapidly with distance than it does for pipelines, it can become an attractive option even in cases where long-distance pipelines would be feasible. Once the distance between producer and market exceeds about 4,000 miles, for example, it would usually be cheaper to transport gas as LNG, and if the alternative were to be a submarine pipeline, the break-even distance would come significantly sooner than this, at probably less than 2,000 miles. (See Figure 4-1.)

With rapidly increasing knowledge of metallurgy and electric welding techniques along with steadily improving methods of corrosion prevention, construction of heavy duty lines up to 56 inches in diameter that are designed to operate at seventy-five atmospheres, is now undertaken, although earlier experiments by the Soviet Union on superlarge diameter pipes of 2.5 meters did not prove successful, mainly due to climatic, handling, and other logistical problems. In relative terms, a pipeline involves high capital expenditure but has a low operating cost. This means that in order to reduce the unit transport cost to a minimum it is desirable to operate at the maximum possible loading throughout the year. Therefore, in designing a pipeline system, due attention must be given to the possibility of meeting winter peak demands either by the provision of sufficient storage for gas supplied during the summer or by some other method of "peak shaving."

Demand for gas is influenced by the weather, domestic habits, industrial working hours, and many other factors. As a result, in industrialized countries with cold or temperate climates, the winter peak offtake may easily be several times greater than the summer trough. Furthermore, daily peaks of even greater disparity occur between, for example, breakfast and evening cooking and heating periods and the early hours of the morning. Gas-supply systems need to deal with those peaks or gas pressure might drop at the appliances of the customers.

Seasonal demand peaks are often hard to deal with, and costly measures may have be employed since no supplier could operate economically with pipelines operating at half capacity or less for part of the year. A typical compromise is to set the pipe capacity at around two-thirds of the maximum offtake. The type of peak shaving employed depends on individual circumstances and, for example, with small gas supply systems with a very limited peak season, capital cost can be minimized by feeding into the system a carefully controlled mixture of liquefied petroleum gases (LPG) and air. On a larger scale of capital and operating cost, there is the possibility of manufacturing gas from petroleum feedstocks and using this for peak shaving. When available, depleted oil or gas fields may be used for storage for peak shaving since it is often possible to retrieve gas pumped into suitable geological formations without major losses. A further alternative is a small-capacity liquefaction plant that could feed LNG into storage for most of the year at atmospheric pressure, so that when the gas is needed for peak shaving purposes the liquid can be fed through a vaporizer and returned to its gaseous state. It may also be possible, and is indeed often necessary, for local gas distributors to negotiate "interruptible" sales of gas to industrial users, whereby gas will be made available at a low price during low-demand periods of the year but will be diverted to other users at peak periods.

Major gas transmission networks now exist in all major consuming markets, and technology is now enabling pipelines to be laid beneath greater depths of water, which has brought a new dimension to the transport of gas. One outstanding example is the trans-Mediterranean pipeline, scheduled to convey up to 12 mrd m^3 per annum of gas from Algeria to Italy, including a 160-kilometer stretch of 20-inch pipeline from Tunisia to Italy at water depths up to 600 meters, and all over the world an increasing proportion of natural gas is being produced offshore and brought to land for processing and distribution. The importance of pipelines as a means of transmission is such that over 95 percent of all the natural gas consumed in the world today has traveled from wellhead to customer entirely through a series of pipelines.

Where, for logistical or political reasons, a pipeline cannot be used, transport of natural gas as LNG may be a successful alternative. The first trial cargo of LNG was shipped from the Gulf of Mexico to the United Kingdom in 1959, and since that time a number of commercial projects have developed so that, today, ocean transport of LNG can be regarded as little more exceptional than the transport of crude oil. The liquid gas is produced by compressing, cooling, and expanding from a high pressure to produce a low-viscosity liquid boiling at about minus 161 degrees centigrade at atmospheric pressure. Although some 15 percent of the heat value of the original may be lost in conversion, the resulting LNG is approximately 600 times more dense than its original gaseous state, so that the capacity of a refrigerated storage tank or ship is increased by the same amount. It may be stored either underground or in well-insulated above-ground tankage lined with aluminum or suitable grades or stainless steel. Since LNG has a specific gravity about half that of crude oil and a calorific value per unit weight of not much more, a particular size of LNG carrier can carry only about half of the heating value that a similarly sized oil tanker can carry. On arrival at the receiving terminal, the LNG will initially be stored in refrigerated tanks and later regasified and fed into the transmission line.

Although LNG still makes up less than 3 percent of the world's gas consumption it is likely that, if the world's remotely situated reserves of natural gas are to play their full part in satisfying energy needs, there will have to be substantial further expansion in the LNG trade. That this is quite possible may be illustrated by the fact that, on one LNG project alone, the thousandth cargo was delivered during 1980 from Brunei to Japan, following an eight-year period of incident-free operation, and today, total contracts for LNG trade amount to almost one million barrels per day of oil equivalent.

Because LNG is nevertheless a costly method of transporting natural gas in view of the sophisticated equipment required and the energy losses, considerable effort has been expended in the search for alternatives. One of the more realistic of these is based on the fact that natural gas can be converted into methanol, a liquid that is presently used in the chemical industry but that can also be used as a low-sulphur fuel or a supplement to or replacement of gasoline for automotive engines. One problem with this method is that the energy loss through methanol conversion and transport, even by normal oil tanker, may be in the order of 35 to 40 percent, compared with the 15 to 20 percent for LNG, and it is generally believed that if a decision is to be made purely on the economics of transport, methanol conversion will not match LNG as an economic means of transporting larger quantities of natural gas. An annual gas flow of some 5 to 10 mrd m^3 will usually, depending on location, be needed to justify an LNG plant, but a large-scale methanol plant only requires about one-tenth of this volume. Thus, if reserves of natural gas are small it may make economic sense for a country to think about entering the international methanol trade, provided it can find suitable markets. It is worth mentioning that countries with a surplus of natural gas may also contemplate exporting it in the form of chemical products, fertilizer, and so forth. In general, economic returns from such routes are thought likely to be less than those resulting from direct sales of gas; therefore, much will depend upon a country's general economic situation and need either for exports of this nature to earn foreign currency or for the greater use of them domestically. At the present time, therefore, pipelines and LNG remain the only proven methods for transmitting substantial quantities of natural gas economically over longer distances.

As already indicated, there are a number of important and individual factors that distinguish the long-distance movement of natural gas from that of oil and that arise from the capital-intensive nature of both the pipeline and LNG business, where the cost of a single project may approach $10 billion. The Alaska Highway Pipeline substantially exceeded that figure. In order to raise finance for such a scheme, it is necessary for the promoters to have reliable outlets backed by firm take-or-pay contracts for the duration of the project, usually twenty years or more. This has led to what has been called the "closed loop" system, with sellers supplying only one buyer or group of buyers with dedicated facilities and ships. There is therefore, unlike oil, little or no spot trade by which buyers or sellers can dispose of surpluses or pick up additional quantities. The international gas trade is therefore rather inflexible and less subject to rapid change than is the oil industry.

As mentioned earlier, one factor arising from the high capital costs is that, for economic reasons, it is necessary for an international supplier to use his facilities at the highest possible level through the year in order to reduce unit costs. Unfortunately, many of the best international markets for natural gas are not of this kind at all, requiring much higher deliveries at peak periods that occur most often during the coldest part of the year. If all gas were to be supplied in this way, the supplier would have to charge very high costs during the off-peak period in order to spread the amortization of his capital, and this would have major implications for the gas distribution and marketing business.

Another key factor that results from the nature of the gas business is that it is difficult to establish an international price for gas in the same way as there is an OPEC price for oil. In the first place, around 85 percent of all gas is consumed in the country of origin, often distributed by state-controlled companies, so that a number of widely different market prices have evolved, depending on the energy policies of individual governments. Where gas is traded internationally, the cost of transporting it is typically between five and ten times higher than that for oil. Thus, transportation forms a major part of the landed cost for LNG or of the border price for pipeline gas. If there were a standard selling price, this would mean a wide range of gas prices in the consuming countries depending on their distance from the supply source and thus the cost of transport. If, on the other hand, an importing country should be strong enough to impose a standard buying price, the value to a geographically more favorably positioned supplier would be much more than that to a less fortunate counterpart. Both buyers and sellers have to understand and appreciate these factors if they are successfully to negotiate the development of international trade in natural gas. Too rigid a stance by either party could well mean that a potential supply of gas does not get traded or that, if it does and an unsatisfactory balance between buyer's and seller's commercial interests is incorporated into a contract, much time and effort may subsequently be expended in renegotiation or in legal disputes.

THE NATURE OF THE GAS MARKETS

A further element of inflexibility in the gas business is related to the nature of the markets that natural gas primarily serves and the fact that buyers of natural gas are nearly always public utilities or natural gas distribution organizations or municipalities. Every country organizes its gas industry to suit its background and political aspirations, and the relative effectiveness of the various ways in which the organization has evolved is probably best judged by the quality of the service given to the end consumer of the gas. In the United States, for example, there is a private enterprise system in which there is real competition between gas producers and gas transmitters. The distributors are given a monopoly right to distribute gas in their franchise area—a practice that is common to all countries—but they have to compete there with private oil, electricity, and coal industries. The resulting competition was a vital part of the remarkable early success story of the American natural gas industry. It subsequently became necessary to regulate the distribution and transmission functions of the industry, but the success of both the methods and the results of that regulation is questionable. The American gas industry is not today in a good financial condition and could have difficulty, as things now stand, to take the actions necessary to ensure customers' future supplies. There are some American distribution companies that distribute both electricity and gas, and some that are municipally owned, but very few of these companies have proved as effective as the private organizations that sell gas only.

In Great Britain, by contrast, the British Gas Corporation is the nationalized undertaking that has the monopoly right to "manufacture," transmit, distribute, and sell natural gas, and fortunately, they compete vigorously with the nationalized electricity and coal industries as well as with the private oil industry. The problems of matching supply and demand in the United Kingdom were appreciated at the time when natural gas was about to be introduced; therefore, some interruptible sales contracts were made for supplying large industrial boilers while pressures to supply gas to power stations were resisted.

In France, Gas de France is the nationalized undertaking having the monopoly rights, but because, for cost cutting reasons, Gas de France and Électricité de France have to share showrooms, meter readers, and some local branch staff, there is little competitive pressure between gas and electricity at the service level. Thus, the end consumer receives less good service. In West Germany, most of the German distributors are municipal distributors of electricity, gas, and water, and they see themselves as existing to provide a service to the public rather than, for example, to actively market gas. The result has been that the rate at which natural gas has penetrated the heating market has been much slower than it could have been, and as a consequence of lower profitability the distributors have not been able to give such good service to their customers as is available in the United States, France, and Great Britain.

It does seem, from analyzing these and other countries' gas industries, that there are considerable benefits to be derived from preserving interfuel competition, and it is also clear that, since circumstances in the energy industries, as elsewhere, change considerably with time, regulations must also change. They must also be fully intelligible, along with the complexities of the gas business as a whole, to those regulators who have to make the changes, often under pressures of special interest groups of all kinds.

However, just because a country's natural gas industry is often a nationalized organization and because the end users include a substantial number of individuals, the inflexibilities mentioned above tend to get embodied in a politicized matrix as well. Votes are very directly involved and threats to housewives' supplies quickly become tantamount to national threats, and this is a factor that has to be taken into account when considering the reliability of imported supplies. But there are many markets other than housewives, and an understanding of the interrelationship between the various sectors is necessary if the effects of pricing at the international level upon the evolution of market shares at the national level are to be fully appreciated.

The domestic and commercial market includes not only use for residential housing but also what are known as "commercial" consumers such as offices, restaurants, hotels, and hospitals, where the energy needs are broadly similar. Typically, these will consist of a demand for cooking and water heating, where volumes are low and insufficient on their own to provide a satisfactory base load. Nevertheless, since the only realistic competitive fuel may often be electricity, the value of gas for these uses may be extremely high. The additional major demand will be for space heating, whether in the form of central heating or "point heating," an example of the latter being gas fires. While this market has pronounced seasonal characteristics that make supply and demand more difficult to balance, it is nevertheless a market where the high qualities and flexibility of natural gas are extremely advantageous. As always, when the competing fuels are either electricity or middle-distillate oil products, the potential value of natural gas is high, and indeed, the domestic space-heating market has been the cornerstone of many of the more successful natural gas marketing operations around the world. Gas utilities would obviously be glad if they could develop a third element in these markets—namely, a summer demand—but this has so far been limited to swimming-pool heating, and considerable research is being applied to develop, for example, a suitable form of gas-fired air conditioning.

There are a number of industries, such as metallurgical, ceramic, glass, and baking, where natural gas can command a particularly good price because its inherent advantages of cleanliness and controllability are useful in the processes employed by those industries. Because this is a regular, year-round, high-value use, it is extremely important to the gas distribution company, and efforts are continually made to expand this category of users by further research and development work. In other industries, such as steel and cement, there is no technical advantage to be gained from the use of natural gas, as most of their needs are for nonpremium energy supplies. Nevertheless, this kind of industry does provide a fairly steady base load, with limited seasonal and hourly variations, and in some countries substantial quantities of natural gas are used in this sector. Because the competing fuels are normally the lower priced energy fuels such as heavy fuel oil or coal, if large volumes of natural gas are used in these markets it is probably because there are plentiful indigenous sources that, together with internal price policies, enable natural gas to be relatively cheap compared to imported fuels. It also may be because gas may be sold to the industry concerned on an "interruptible" contract that may state for example that, under specified terms of notice, the gas distributor may stop supplying the user with gas when gas demand for other sectors is at its peak. This is a useful technique for leveling out gas demand through the year, but it also means that, in order to provide sufficient incentive for the industry to secure alternative fuel during periods of interruption, the price of gas will have to be lower than that of competing fuels. Interruptible supplies take on added significance when discussions turn to security-of-supply issues involving imports of natural gas from countries that may be of uncertain reliability as long-term suppliers. For example, should there be, for political, technical, or climatic reasons, an interruption of imports, supplies of other gas–from domestic or manufactured or other import sources–would have to be directed to those markets most vulnerable to cut-offs. The purchasers of interruptible contracts would obviously be the least vulnerable, in view of their existing alternative arrangements. They could therefore be interrupted without ceremony, and their gas could be diverted to the markets most adversely affected.

Natural gas is used for electricity generation in some countries, and the same competitive aspects apply as in the case of industrial under-boiler use. However, a particularly significant outlet has grown up in Japan, where, air pollution legislation is extremely severe and a substantial proportion of the base-load power generation, in densely populated cities like Tokyo and Osaka, is based on imported LNG. In other major gas markets, for instance the United States, the Soviet Union, and northwest Europe, this market sector was in the early 1980s still an important user of natural gas, but if prices continue to rise relative to other fuels, as has been the case in recent years, consumption is likely to fall. On the other hand, countries such as the Gulf states, where natural gas is plentiful and where a substantial proportion of energy demand is for electricity to power air-conditioning units, it is common for natural gas to be used as the main fuel, freeing more valuable crude oil for export.

SOME ASPECTS OF PRICING

Following this necessarily cursory look at the main market sectors for natural gas, it is appropriate to move on to consider some of the aspects of natural gas pricing. It has already been pointed out that there is no such thing as a world price for this particular energy form in view of the historical development of the various major gas markets and as a result of the geographical disparity between production and consumption areas coupled with the high "cost of distance." It would therefore be misleading to pretend that what follows is a comprehensive account of the myriad factors that influence the pricing of internationally traded gas, since every case must be considered in light of geographical locations, market mixes, the degree and nature of government involvement, and so on. Indeed, the pricing of any particular major gas deal will ultimately be resolved by negotiation between the principal parties, subject to such constraints of a governmental or similar nature that may apply in any particular instance. The most that can be attempted here is to present some general wisdom on some of the trends and developments in gas pricing in the hope that this will serve to clarify further the differences and similarities between the international oil and natural gas industries.

Reference has been made to the disparity between international gas prices and indigenous gas prices, but it is interesting to consider what has been the trend in delivered prices of internationally traded gas in the major global markets of the United States, Western Europe, and Japan. In the United States, the average price for indigenous gas was, in 1981, little more than $1.50/MMBtu, with prices ranging from $0.30 to $3 ex-field for regulated gas and as high as $10 for small quantities of unregulated gas. The average low price for indigenous gas enables the relatively small quantities of high-cost unregulated gas and imported gas, which collectively amount to less than 10 percent of total consumption, to be "rolled in," that is to say, mingled with the cheaper gas, the price to the end users being averaged out accordingly. Gas imported into the United States by pipeline from Canada and Mexico was at this time priced at just under $5/MMBtu, while the small quantities of Algerian LNG that are still being imported by Distrigas were delivered c.i.f. (cost, insurance, and freight) at about $5.10, which, after regasification, brings the price of this gas to nearly $6. Additional Algerian gas was not, in the early 1980s, being imported into the United States largely because, for the El Paso scheme on which discussions started several years earlier, Algeria was still insisting upon a base f.o.b. price of $3.70/MMBtu, indexed after April 1, 1981 as to 100 percent with increases in a basket of five crude oil prices. The U.S. government was not prepared to go beyond 70 percent indexation although, had agreement been reached, it would have resulted in an ex-terminal price of $5.61. For another Algerian gas scheme, the Trunkline contract in which approved prices were fully indexed with Nos. 2 and 6 oils, the delivered price, after regasification, would have been more than $7.40, based on an Algerian f.o.b. asking price in excess of the $4.12/MMBtu as provided for in the original contract.

Algerian insistence both on raising f.o.b. prices after contracts had been approved, and on f.o.b. prices that would result in netbacks in line with those Algeria received for its oil exports,[1] resulted in the postponement of Trunkline deliveries and the cancellation of the El Paso project. The development provides an excellent example of how the flow of internationally traded natural gas can be impeded or prevented by one or the other partner in an export scheme upsetting the stability that is so essential for a long-term high-cost project of this nature. Whatever may have been the justification for Algeria's attempt to raise export prices to crude oil parity levels, it was the decision unilaterally to change horses in midstream that proved a severe setback to what otherwise looked like a promising beginning to an East-West transatlantic expansion of the international trade in natural gas. The cause of imported gas into the United States was in no way helped by these events, which also tarnished the image of LNG generally in that they demonstrated how unreliable it might be. It was perhaps fortunate for the consumers in the United States that there was developing in the early 1980s a temporary surplus of indigenous gas deliverability. However, El Paso must have suffered severely commercially, and Algeria must also have lost substantially, not only from the point of view of export revenues but also in terms of reputation and through the underutilization of its export plant facilities. Perhaps it was a lesson that had to be experienced by someone sooner rather than later in the evolution of the international trade in LNG, but it is to be hoped that all potential or existing suppliers and purchasers of natural gas will have benefited from the U.S. and Algerian experience to the extent of appreciating the fundamental necessity for adherence to contract terms once these have been negotiated and agreed upon.

Pricing in Western Europe is always a complicated picture since there exists there a complex mix of quite separate national markets. Over recent years, the average delivered price for international sup-

1. Dr. M. Belguedj, "The Supply of Natural Gas in Western Europe" (Paper presented at the European Gas Conference, Oslo, Norway, May 25th, 1981).

plies had been rising steadily, and in mid-1981 existing contracts were being supplied at around $4/MMBtu. It was becoming apparent that, toward the end of that year, the price of existing supplies might have to be negotiated upward (to $5) and that new gas deals would be likely to be concluded at nearer $6 than $5/MMBtu, particularly since Algeria, which was waiting for another major Soviet export deal to be concluded with West Germany and France, was militantly seeking crude oil parity for its exports, which would have indicated a price over $6. However, at about that time, a substantial reduction in West European overall energy consumption, as a result of the general economic recession, was becoming apparent. This, together with downward revisions in natural gas demand estimates and total opposition to the Yamburg-Urengoi pipeline on the part of the Reagan administration, convinced the Soviet gas exporters that they would never secure a deal with Western Europe if they adhered to the high prices—$6/MMBtu—they originally asked. It was not until asking prices had come down to under $5 that West Germany and France were persuaded, in the autumn of 1981, to sign up for purchases of, respectively, 10.5 and 8 mrd m^3 per annum over twenty-five years. Further contracts seemed likely to be signed at around this level by the Italian, Belgian, Austrian, and Swiss importers, but there were considerable political overtones to the new Soviet export deal that could be considered somewhat distorting to the free play of market forces. This issue will be discussed further in a later chapter.

Including indigenous gas, West European wholesale gas prices were, in 1981, spanning a range of from less than $1 to over $5/MMBtu in a few isolated cases, although the range for existing international gas supplies had been narrower, averaging around the $4/MMBtu mark, depending on exchange rate fluctuations. In the main, existing international gas supplies are indexed up to 100 percent with either oil product prices, in particular fuel oil, or with crude oil, or some combination thereof, with varying time lags in the application of such indexation provisions. Oil product prices in Western Europe in the second half of 1981 were extremely volatile and subject to seasonal swings, but there were signs that wholesale international gas prices were headed toward the $5/MMBtu mark—a level that most gas utilities were expected to be able to accommodate even though their market mix contained a substantial volume of low-value, under-boiler business. In the event that international gas would have approached or exceeded $6/MMBtu, or close to crude oil parity, this would have substantially exceeded the then-current Rotterdam prices for fifty-fifty gas oil-low sulphur fuel oil by $2 in the case of heavy fuel oil.

One deal that caused some West European gas utilities considerable concern was that between Distrigas of Belgium, and Algeria, in which the base f.o.b. price of $4.80, as of January 1, 1981, was indexed 100 percent with quarterly changes in the government selling prices of certain crude oils. Both Algeria and Libya were seeking to upgrade their existing contracts with France and Italy, and it was clear that, before the signing of the Soviet-West German deal toward the end of the year, pressures on prices were definitely upward. If indigenous gas prices would have moved upward in a complementary fashion the market would have found, say, an additional dollar on the price very difficult to absorb. Indeed, at $6/MMBtu, growth in West European demand for natural gas would probably have ceased altogether for the time being, and it might even have declined had the price gone any higher.

In Japan there were sharp increases in the price of LNG imports from Alaska, Abu Dhabi, and Brunei during 1980, when prices more or less caught up with that of Indonesian LNG, which had been indexed as to 90 percent with crude oil. In the autumn of that year the Abu Dhabi LNG even overtook the average delivered price of imported crude oils because it had been indexed with the high-priced Murban crude. From early 1981, the range of delivered prices for imported LNG widened again to the point where, in July of that year the cheapest, from Indonesia, was at $5.11 while that from Abu Dhabi was $6.62/MMBtu. During the period from June 1980 to July 1981, prices of other low- or zero-sulphur fuels—such as LPG and Naphtha—fell from such high levels as $7.40 and $7.95 to $6.12 and $7.11/MMBtu respectively, and since the average crude price rose from $5 in early 1980 to $6.40/MMBtu, LNG prices became increasingly expensive in relative terms. The Japanese gas pricing scene had become no longer as consistent as it had been prior to 1980, and Japanese buyers were obviously becoming increasingly uncomfortable with the divergence of LNG prices and with the linking of such prices with crude oil at a time when certain oil product prices were eroding. A number of schemes for importing new LNG—from Indonesia, Malaysia, and Australia—were under construction or advanced negotiation at the time. It was questionable whether the short-term fluctuations were unduly coloring the underlying longer term realities since Japan, alone among the major gas markets, has virtually no indigenous energy of any consequence and remains uniquely dependent on international supplies. It therefore faces different problems and challenges for gas than do Western Europe or the United States, which still enjoy a large base of relatively low-priced indigenous gas. It was indeed still unclear in mid-1982 whether Japan was a buyer's or seller's market.

CRUDE OIL PARITY?

It can be seen that the concept of crude oil parity pricing on a thermal basis is central to many of the issues surrounding the debate on internationally traded natural gas supplies, and it is appropriate at this point to consider both sides of the question. In the view of those producers and sellers who are striving to raise export prices, since energy generally has become scarcer and since natural gas is at least as valuable as the oil it replaces, crude oil parity is a realistic pricing aim. Natural gas, they point out, competes in the domestic and commercial sectors with gas oil and electricity and should be priced accordingly, and the fact that some consumers choose to use natural gas as low-value under-boiler fuel should not be a criterion for determining its price. They point out further that because natural gas is more expensive to transport than oil, and because some is lost during liequefaction and compression, they need adequate and guaranteed returns after paying such costs and claim that they cannot afford to subsidize consumers. Producers believe that LNG should be priced on a different basis to gaseous gas piped over short distances because of the different costs involved and hence the difference in netbacks, and they consider that gas should in any case not be priced to reflect today's market mix but at its full value, so as to force it out of low value into higher value uses. Gas, they emphasize, is a high-quality fuel and should move quickly in the marketplace to a price level nearer to electricity or to a break-even level with synthetic natural gas produced from coal, so that pricing at parity with crude oil is simply a first step in this direction. Furthermore, they consider that the buyers already have the greater part of their distribution structures in place and paid for, while, for new supplies, it is the seller who has to make substantial new investments before any new revenue is generated.

As against all these points, the buyers argue that natural gas is not a substitute for crude oil itself, but only for a number of oil products of differing prices that happen to be derived from crude oil. Oil is likely to be increasingly used for transport and chemical feedstock purposes and may even become largely decoupled from the heating market, which is the main outlet for gas, so that crude oil prices may therefore become less representative of the price of heating fuels. The price of gas should be based on its value in the market and not on any arbitrary crude oil prices, which have often been changed at the whim of host governments in the past. Buyers also contend that end consumers make the final choice and that natural gas must therefore compete at the consuming point after allowance has been made for any quality premiums and other costs involved. While a premium for supply cost and security of supply may be justified, gas priced f.o.b. or ex-field at crude oil parity is too expensive to sell in the market because of the high transportation costs, and in any case the reliability of supply for imported foreign gas is becoming increasingly suspect.

Whatever the strengths or weaknesses of the arguments put forward for or against crude oil parity pricing for internationally traded natural gas, it is possible that it will only be a matter of time, albeit years rather than months, before the concept of crude oil parity, already established on a delivered basis in Japan, becomes the norm for the delivered price of imported gas in Western Europe and the United States. It is difficult to come up with another realistic and viable pricing alternative, at least in the short to medium term, that is likely to encourage producer governments to develop new export projects. Looking even further ahead, even a price approaching crude oil parity may become insufficient to stimulate the development of yet further new supplies.

The mystique and secrecy that surrounded many gas prices in the past is rapidly evaporating, and while some degree of confidentiality will probably always be necessary on certain contractual aspects that are not matters of public record, it could turn out that the greater degree of openness now emerging on this sensitive subject will lead to a better understanding between sellers and buyers of each other's aspirations and constraints, to the mutual benefit of both. This chapter has hopefully demonstrated that, while natural gas and crude oil are both internationally traded hydrocarbon energy forms, there are considerable differences between them that profoundly influence the roles they can play, and it seems that these differences will never be totally reconciled even if a greater approximation between oil and gas prices evolves over the next few years. But, like oil, natural gas is going to be required on a global basis in ever-growing quantities over the next twenty years and numerous large investments will have to be made to ensure its availability. Since domestic reserves in all major markets except the Soviet Union are limited or declining, the volume that must be internationally traded will have to grow substantially, and a great many physical, commercial, and political problems are going to have to be solved in the process. This particularly applies in the case of Western Europe, which will now be examined in the context of natural gas supplies and requirements up to the end of this century.

5 WESTERN EUROPE

The chief reason that this book deals particularly with the future of natural gas supply to Western Europe is that Europe, alone among the continents of the world, is afflicted by the great ideological divide between East and West. On one side of that divide is located some 35 percent of the world's proven reserves of natural gas; on the other side are markets representing over 15 percent of global demand for this hydrocarbon. Whatever political or ideological barriers there may be between Western and Eastern Europe, they share the same land mass, and natural gas, along with other internationally traded commodities, and people can move from one part to the other without the need for a costly and time-consuming sea journey. The strategic and security implications of this geographical relationship are obvious, but while Western Europe is within economically practical reach of a number of other natural gas supply sources, it would be difficult to ignore the logic behind a substantial share of West European demand being met by deliveries from the huge reserves situated just beyond and to the south of the Ural Mountains.

Few people could have expected, as little as twenty years ago, that the question of the East-to-West flow of natural gas in Europe would become, during the 1980s, of major significance for East-West relations as a whole, and for relations within the Western Alliance in particular. Indeed, few people in Western Europe had any reason to be aware of natural gas at all in the 1950s, for there was no international trade in the moderate amounts then being produced from small reservoirs in Italy, France, the Netherlands, Austria, West Germany, and the United Kingdom, and the gas industry in general was under severe competition from low-priced imported oil, which had rapidly replaced coal and manufactured gas as the major growth fuel after the World War II. Local West European gas markets were, by the end of the 1950s, receiving some 12 mrd m^3 of natural gas, mainly in Italy and France, a similar but declining volume of gas manufactured from coal, and about 2.5 mrd m^3 manufactured by newly developed processes from oil-based feedstocks. Total gas met about 4 percent of primary energy demand in 1960 and natural gas therefore represented less than 2 percent of supplies. The Soviet Union had only recently started to produce significant quantities of natural gas, all for domestic consumption.

GAS FROM HOLLAND

The rather gloomy outlook for the West European gas industry was transformed by the discovery in 1959 of a huge natural gas field in the Groningen province of the Netherlands, with reserves subsequently estimated at over 1.5 trillion cubic meters. This was nonassociated gas, and the Groningen field is today among the top six of the world's largest known nonassociated gas fields. Not only was this major discovery located on land, in an industrialized country with high living standards where gas could be readily utilized, but it was within close proximity to the major industrial conurbations of Belgium, Germany, and France and within a commercially feasible distance of Southern Europe as well. The economic worth of and scope for Groningen gas was thus potentially considerably greater than for many similar large gas accumulations in various countries in the Middle East, Africa, and the Soviet Union which tended to be remote from major gas consumption markets. The quality of Groningen gas is actually somewhat different from that of many major natural gas fields in that its calorific value is only 8,400 kcal/m^3 compared with values for other supplies close to or in excess of 9,400. This, however, did not present any major barriers to its utilization and it rapidly became the basis for a spectacular increase in natural gas consumption in many West European countries. The

Netherlands government decided that the Groningen gas would be sold on a commercial basis consistent with the greatest possible benefit to the national economy and that the gas should be introduced as smoothly as possible, without undue dislocation of the existing energy market. The first delivery of Groningen gas for domestic distribution occurred in 1963.[1] Small quantities began to be exported in this same year to West Germany but, during the early and mid-1960s, a series of export deals was concluded between the Dutch gas export organization, NAM/Gas Export—a part-government/part-private industry—and buyers in Belgium, France, and West Germany, and new gas transmission pipeline systems were built to implement these deals. Export prices at the border were mainly determined by the prices for heavy fuel oil ex-Rotterdam, the cost of transport to the export markets and the conditions prevailing in those markets. From 5 million m^3 in 1963, exports rose rapidly to 11 mrd m^3 in 1970, and domestic consumption in the Netherlands increased from 806 million to over 20 mrd m^3 in the same period. In the second half of the 1960s and early 1970s, further export contracts were concluded with West German buyers and with Swiss and Italian gas interests. By 1976, a peak of export volumes occurred at 50.6 mrd m^3, and with domestic use requiring nearly 44 mrd m^3, that year witnessed the maximum Groningen sales of 94.3 mrd m^3. Most of the natural gas export contracts that were concluded in the 1960s were at fixed border prices, in a period when energy prices were falling and energy as a whole was in plentiful supply, and they stipulated that periodic price adjustments could be negotiated. Such a condition is what one might expect for a contract that had a life span of twenty to twenty-five years, for it would not have been reasonable to expect either contracting party to have to remain rigidly bound to an immutable contract over such a period. Indeed, during and immediately after the oil price rise shocks of 1973 and 1974, the previous system of more or less fixed Dutch border prices was abandoned and more flexible pricing mechanisms linked to oil prices were introduced by common agreement. Many of the export customers for Dutch gas ultimately began to receive increasing volumes of natural gas from other sources, and during 1976 and 1977 as a result of an extensive round of negotiations lasting over a year, export border prices were readjusted in exchange for greater offtake flexibility, coupled with extensions to the contract period although with no

1. Malcolm W.H. Peebles, *Evolution of the Gas Industry* (London: Macmillan, 1980).

increase in the total volumes to be supplied over the revised and extended contract period.

As a consequence partly of these latter adjustments, export volumes from Holland leveled off during the late 1970s and early 1980s, but it also meant that export customers could receive Dutch natural gas over a longer period of time, and their supplies will not terminate so abruptly as they might otherwise have done. A further reason for the cessation of Dutch export increases was the fact that toward the end of the 1970s, economic activity in Western Europe started to become depressed, and many countries introduced energy conservation policies that involved, for example, a gradual reduction in the use of natural gas for power generation purposes. Unless major new discoveries are made, no new export contract for Dutch natural gas will be entered into, and export volumes will gradually decline for the next two decades. Thus, by the year 2000, virtually all the Dutch export gas will have been replaced by other supplies, and Groningen production will be retained entirely for consumption in the Netherlands.

So the end of this remarkable era in the development of the natural gas industry in Western Europe is in sight, but the availability of large quantities of cheap and reliable Dutch gas, conveniently situated near large markets, has shown what can be achieved in such circumstances in a comparatively short time, and it is difficult to deny the fact that the Dutch developed the particular natural energy resource, which they were lucky enough to find under their territory, in the most competent economic, commercial, and technical manner for their own and the common good. This must contain a valuable lesson on how the interests and objectives of governments and private industry can be harmonized in a way that does not detract from the overall public interest.

Dutch gas was the only source of natural gas supply for Belgium until 1977, and it is still, pending the delivery of additional Soviet gas in the mid-1980s, the largest single component of supply for the West German gas market. It is also the predominant source of gas supply for northern France and has become an important factor in supplies for Italy. Without Dutch gas, it is possible that some of the European gas markets might not have survived the severe competition from cheap and abundant oil in the 1960s, and it is worth noting that Dutch gas as a whole constituted nearly half the total consumption of natural gas on the West European continent during the 1970s and represented some 700,000 barrels per day of hydrocarbon energy that did not have to be imported from non-European sources during a particularly turbulent time for international energy supply. If large reserves of natural gas had not been discovered in the British sector of the North Sea in the mid- to late 1960s, it is very possible that the United Kingdom would have become another substantial customer for Dutch gas and would now, together with all the other customers, be having to take appropriate steps to see that supplies from non-Dutch sources would be available in sufficient quantities to meet their demands for gas in the marketplace.

While this section has concentrated upon the impact that the discovery and export of Dutch natural gas had upon the development of the European gas industry as a whole, it is worth mentioning that the Dutch economy benefited greatly, not only because of the savings evolving out of not having to import so much other energy, but also because of the encouragement that the availability of reliable and plentiful indigenous energy gave to domestic investment. Revenues from the export of Dutch gas rapidly became an important element in the State budget, and they were put to good use from the second half of the 1960s in the financing of substantial social and welfare programs. By the end of the 1970s, around 90 percent of all homes in Holland were heated by natural gas, which also provided over 50 percent of total Dutch primary energy requirements, a percentage not matched in any other industrialized country in the world to date. Remaining recoverable reserves of Dutch gas are still considered to be in excess of 1,500 mrd m^3, and it is not surprising that the Dutch should have decided that most of this will be consumed in the Netherlands. It should also be remembered that, for many years to come, the Groningen reserve exists as a significant potential security back-up element for West European natural gas consumers as a whole, and it is perhaps timely to contemplate both what measures might be appropriate to ensure that the strategic reserve is not depleted beyond a certain level and how it might be replenished by imports from non-European or Norwegian sources, provided agreement can be reached between all interested parties on how any such developments might be financed.

GAS FROM THE NORTH SEA

The geographical location and the geological structure of the Groningen natural gas deposit, when compared with geological evidence from the eastern regions of the United Kingdom, indicated that further hydrocarbon reserves might exist beneath the North Sea and nonassociated natural gas was indeed discovered in the West Sole field in the southern basin of the U.K. sector in October 1965. By the end of the 1960s it was becoming apparent that sufficient natural gas might exist there to allow all domestic requirements for gas to be met from this new source. Previously, the United Kingdom's coal-based manufactured gas industry, which had been nationalized in 1949 with the creation of the Gas Council, was facing problems of declining domestic demand and increasing raw-material costs, and by the early 1960s it had become clear that, with new and economic processes for the transformation of oil becoming available, the days of coal-based gas were numbered. The transition of coal to oil feedstocks was accomplished rather smoothly, following government support for the move in 1965, and gas became a growth business once again. The Gas Council had by this time been experimenting with the possibility of importing liquified natural gas (LNG), originally from Louisiana, and a fifteen-year contract was signed with Algeria in 1961 for the delivery of about 100 million cubic feet per day, the first deliveries of which, in 1964, marked the start of the world's first commercial international LNG project. The discovery of substantial quantities of natural gas under the North Sea in the second half of the 1960s raised the question of how the exploration and production effort could be encouraged under a system where, according to the terms of the Continental Shelf Act of 1964, any natural gas discovered in the U.K. offshore sector had to be offered to the Gas Council at "a reasonable price," which was not defined. There was therefore doubt on the part of those private companies, which had acquired licenses to explore, as to what the rewards might be, and they only knew for certain that they would not be able to sell whatever gas they might find to any other buyer until such time as the U.K.'s requirement for gas was fully satisfied. However, a considerable amount of exploration was carried out and a number of supply contracts were signed with the Gas Council, albeit at prices that were considered low by international standards. During the 1970s, with the discovery of associated natural gas in more northerly parts of the North Sea, and with the changing economic and global energy supply situations, there were improvements in the terms available to producers, but natural gas was still sold to U.K. customers at prices significantly below those of competing fuels, and this undoubtedly encouraged a more rapid conversion to gas than would otherwise have been the case. Over 13 million customers who were on manufactured gas in 1967/68 had been converted to natural gas ten years later, and over 13.5 million tons of coal and nearly 5 million tons of oil used for the manufacture of gas in 1967/68 had been reduced to 8,000 and 92,000 tons respectively by 1976/77. A million new gas customers, mainly in the domestic sector, were added during this period, with the average consumption per domestic customer more than doubling from 211 to 453 therms per year. The number of industrial customers declined by about 14 percent over the same time, but sales to this market sector increased from 915 to 6,107 million therms, with the largest single increase being in respect of sales to the chemical industry.

The situation regarding supply and demand for natural gas in the United Kingdom at the end of the 1970s was one that, as a result of supply contracts from the southern and northern U.K. sectors of the North Sea as well as some from the Norwegian sector, called for considerable coordination. The British Gas Corporation was created in 1973 and charged with developing and maintaining an efficient, coordinated, and economical system of gas supply to Great Britain and satisfying, so far as it is economical to do so, all reasonable demands for gas in Great Britain. It had sufficient supply availability under contract or in prospect to meet the expected growth in demand of the public distribution market through at least 1990. This market is expected to grow throughout the 1980s, but even so, there seemed to be supply availability to spare, and this could have been expected to enable British Gas to compete with oil and coal in the under-boiler and chemical feedstock markets.

What effect a substantial fall in oil prices, coupled with reduced availability of associated gas resulting, for example, from future oil production cutbacks, will have upon the gas industry in Britain is not yet clear, but it should be mentioned that, in the early 1980s, prices of natural gas to consumers were being allowed to rise at a rate 10 percent higher than the inflation rate, bringing them more into line with competing fuels. A further significant new development was the curbing in 1981 of British Gas's monopoly power so that producers could now offer directly to major customers. Furthermore, the British government indicated in early 1982 that it could consider the possibility of natural gas from the U.K. sector of the North Sea being made available for export to the European continent, provided it could be demonstrated that sufficient proven reserves existed surplus to those required to ensure supplies for indigenous markets. Paradoxically, although there appears to be more than enough natural gas overall to meet U.K. requirements for many years, with prospects for additional discoveries in the unexplored waters not only of the North Sea but also of the Irish Sea and the Western Approaches, it may become expedient for increased though still fairly small quantities of LNG to be acquired, toward the end of the 1980s, for logistical and peak-shaving reasons. On the whole, however, and taking into account Norwegian deliveries, the United Kingdom seems likely to have sufficient supply availability until about the late 1980s. Thereafter it will probably need to import increasing quantities of gas from Norway or from sources further afield. Depending on the sources involved, this could have an impact upon the supply and demand situation of the rest of Western Europe. Of course, if the natural gas pipeline systems in the U.K. sector of the North Sea were to be, in some way, linked with the continental system, permitting exportable surpluses of natural gas on a short-term or seasonal basis to flow either way, this could have significant implications for the overall security and flexibility of Western Europe's supply situation.

Although there are a great many aspects that, for reasons of space, have not been afforded much attention in this chapter, the developments of Dutch onshore and U.K. offshore natural gas have been dealt with more fully than the development of the gas industries in other West European countries will be. Not only are they, for the most part, independent of each other but they are significant for a number of different reasons, and they illustrate a number of different points indicative of where the natural gas industry of Western Europe now stands after the first two decades of development. It is appropriate at this point to consider briefly the situation in respect of natural gas developments in the Norwegian sector of the North Sea, since the availability or otherwise potential volumes of Norwegian gas is of considerable significance to the future supply security of continental natural gas.

Several associated and nonassociated gas fields lie on or close to the median line boundary of the North Sea between the United Kingdom and Norway, one of which, the Frigg field, came on stream in the autumn of 1977. Since Norway has no domestic gas markets, the gas from Frigg was purchased by British Gas, having been competed for by other West European buyers. The existing proven Norwegian reserves south of the sixty-second parallel are indeed already far more than Norway is likely to be able to utilize itself over a very long period, but geological evidence indicates that a great deal more gas may be discovered both in the Norwegian North Sea sector and in waters further to the north.The policies adopted by the Norwegian government for the exploitation of its hydrocarbon resources will of course determine the extent to which other West European countries will have access to the Norwegian gas. So far, Norway has adopted a policy of "moderate depletion," which is applied on a national basis, in contrast to the U.K. policy, which is applicable to individual fields. Toward the end of the 1970s, the Norwegian government decided to aim at an annual hydrocarbon-production level of a maximum of 90 million tons of oil equivalent, as an illustration of a moderate tempo for the development of their North Sea sector, with 90 million tons being about ten times Norway's own consumption at that time. The major upheavals in the world energy markets caused by the oil price rises in 1979 did not cause any major change in the government's approach, and the policy was not apparently altered by the government that came to office in 1981. An important consideration was the risk of damage to the social fabric of the nation that might result from the massive inflow of revenues that would assuredly accompany development of all new natural gas discoveries as soon as the technical and economic bases for that development were established. It was feared that if free rein were given to all those interested in securing supplies of Norwegian gas as quickly as possible, the ceiling of 90 million tons oil equivalent might be exceeded. There was also considerable pressure against the too-rapid development of the offshore deposits, from conservationist and, particularly, fishing industry groups who feared the deterioration of the natural environment that it was felt would inevitably ensue. It is of interest to note that the relevant production figure for 1980 was around 50 million tons of oil equivalent, more or less equally divided between oil and natural gas, and that this is expected to increase to around 55 million tons in 1985, so there is quite some way to go before the government limit is reached. Since there are formidable technical and climatic problems to be overcome in developing the North Sea deposits, it is unlikely in any case that substantial schemes for Norwegian deliveries in addition to those already in operation or under construction can physically come on stream before the early to mid-1990s. However, it is also possible that growing domestic requirements for revenues for social development schemes, or a reappraisal of West European energy security issues, could cause the Norwegian government to revise the ceiling figure upward.

At present, spare capacity exists in the pipeline joining the southernmost Norwegian North Sea gasfields in the Ekofisk area with the Continent at Emden in West Germany. There are plans to link more northerly fields such as Statfjord, Heimdal, and Sleipner with Ekofisk and thus with the Continent, but additional gathering and pipeline capacity will be needed if other significant finds, which have yet to be fully proved up, such as North Sea block 31/2, the Troll field, are to be exploited. Schemes were under discussion in 1982 for plans to pipe some of any future Norwegian discoveries, not subsequently purchased and collected by the United Kingdom, initially to the Norwegian shore for treatment and extraction of natural gas liquids. The Norwegians were also studying with the Swedes the possibility of ultimately piping gas through to the Continent via Sweden, which could also be interested in purchases. While much depends upon the policies of the Norwegian government, these may yet be influenced by the continental European buyers and the nature of the proposals that they are prepared to make. This raises the difficult question of how any new Norwegian North Sea developments are to be financed if they are not developed solely for commercial reasons. That is to say, if other governments consider that the availability of reliable supplies of Norwegian natural gas are worth paying a "security premium" for to reduce, for example, the risks associated with the envisaged growing dependency of some West European gas-importing nations upon supplies from the Soviet Union, then who will pay for them and how will they be mobilized? Before leaving this section it should be noted that commercial offshore discoveries of natural gas have been made in the Dutch and Danish sectors of the North Sea, although exploration efforts in the West German sector have not been very successful so far.

WEST EUROPE'S RESERVES

Before attempting to look at how the West European gas industry might evolve over the next two decades it is necessary to examine in some statistical detail the reserves, supply, and demand situations that existed in West European countries at the end of the 1970s

Table 5-1. Western Europe's Proven Natural Gas Reserves, January 1, 1982.

	$10^9 ft^3$[a]	*$10^9 m^3$*
Austria	345	10
Belgium/Luxemburg	—	—
Denmark	2,000	56
Finland	—	—
France	2,720	77
West Germany	6,180	175
Greece	3,800	107
Ireland	860	24
Italy	3,650	103
Netherlands	55,700	1,577
Norway	49,370	1,398
Portugal	—	—
Spain	25	1
Sweden	—	—
Switzerland	—	—
Turkey	550	16
UK	26,000	736
	151,200	4,464

a. Source: "Worldwide Oil and Gas at a Glance," *Oil and Gas Journal* 79, no. 52 (December 28, 1981): 86.

and beginning of the 1980s. The overall figures to be presented will include those for the United Kingdom, except where stated to the contrary, and will be equated to a calorific value for natural gas of 9,400 kcal/m³. If we consider, initially, only proven reserves of both associated and nonassociated natural gas in Western Europe—that is, those reserves that have been quantified with a reasonable degree of certainty and that can be commercially exploited with existing technology—then in 1980 there were some 4.5 trillion m³, of which over three-quarters were located in the Netherlands, Norway, and the United Kingdom. (See Table 5-1.) To put this in perspective, 4.5 trillion m³ is about half the proven reserves of North America, less than 15 percent of those in the Soviet Union, and under 6 percent of world reserves. But more important than the absolute size of the natural gas resource base is the ability to bring, at a cost that is acceptable and competitive with alternative energies in the marketplace, remotely located gas to those markets that need it while at the same time offering producer governments a sufficiently attractive reward to encourage the exploitation of the resources. It is, of course, a very expensive and technically difficult operation to exploit the reserves under the North Sea, but it is not only the severe climatic and marine environments that have to be mastered there. The gas fields so far located are by no means all large or situated conveniently close to one another; indeed, there are a great many small fields, in widely scattered locations, which it may never be commercially practical to produce. Furthermore, the geological conditions in some of the larger fields, particularly on the Norwegian side, are complicated, and while the gas-bearing strata may be fairly close to the sea bottom they are rather thin, and they will probably require a high degree of lateral drilling and a wide, costly gathering system.

When assessing the future availability of indigenous natural gas, the size of known West European reserves is clearly less important than their accessibility and whether or not the producer countries are willing to produce them to a degree in excess of what they will require for their own use. Of course, there must also be a willingness on the part of consumer countries to rely upon the supply sources in question, and in the absence of either of these conditions the gas resources may have little if any value in the overall future supply-demand equation. At current consumption rates, West Europe's proven reserves would last about twenty years if they were all exploited and if no gas was imported from non-European sources. However, this is not a realistic approach to the situation since not only is a large and growing volume of natural gas imported, but there is also little prospect of a substantial proportion of the indigenous proven reserves in some countries actually being used during the next twenty years for both technical/commercial and political/sociological reasons. The availability and price of alternative energy sources and the time factor are also major elements in determining the extent to which West European natural gas reserves will be exploited, and in recent years the new dimensions of supply security and strategic reserves have come to the fore as Western Europe's heavy dependence upon imported energy, at a time of turbulent global energy supply, has been increasingly recognized as a matter of considerable political concern. The question is raised of whether or not a country lucky enough to have substantial indigenous natural gas reserves should conserve them as much as possible for its own future domestic use and possibly for the use of its allies in an emergency, or whether it should use them or export them while the market is good, reaping the benefits of reduced imports of high-cost alternative energies now in the hope that the international energy supply situation will somehow get better.

SUPPLY AND DEMAND

It is obvious that it is not particularly useful to look at Western Europe as a whole when considering the natural gas reserve, supply, and demand situations since there are more than a dozen countries involved, in widely differing geographical, political, and energy supply situations. Therefore, it is necessary to look at some of the individual countries and attempt to determine which have a greater or lesser significance for the future of natural gas supply for the Continent as a whole. In the case of the United Kingdom, ultimately recoverable or probable North Sea reserves remaining in 1980 were estimated to be slightly in excess of 1 trillion m^3, although proven reserves were probably closer to 700 mrd m^3, while production was running at 45.5 mrd m^3, giving a proven reserves-to-production ratio of about 16 to 1. The remaining proven reserves in the Netherlands were some 1.6 trillion m^3 and annual production there was 94 mrd m^3 so that, at that rate, there was a reserves-to-production ratio of 18 to 1. But production is declining steadily, and when, by the end of this century, all exports will have virtually ceased, if local consumption does not rise significantly above its 1980 level of about 34 mrd m^3, there could still remain some twenty to twenty-five years of consumption, assuming that indigenous supply is not augmented by imports at some stage. Norwegian proven reserves in 1980 were in excess of 800 mrd m^3, and substantial additional reserves are likely to be proved up as a result of recent discoveries in the block 31/2 area and elsewhere. The 1980 production level of some 31 mrd m^3 would have indicated a reserves-to-production ratio of 26 to 1 but there is no doubt that the figure can be significantly higher, and at present there is no domestic consumption.

As for some of the other West European countries, West Germany's reserves in 1980 were around 190 mrd m^3 and it was producing 16.5 mrd m^3 so that indigenous reserves would last, at this rate, another twelve years. However, demand was running at about 52 mrd m³ so that two-thirds of consumption was already being imported, and the share of imports will steadily increase in coming years. Since West Germany is one of the key importing countries in Western Europe, its natural gas situation needs to be described in rather more detail. Natural gas consumption in 1980 actually fell by 2.6 percent against 1979 and an even larger reduction was expected in 1981 in view of a downward trend in both the level of economic activity and of energy consumption overall. In 1980, natural gas accounted for 16.3 percent of primary energy consumption, and of the 55 mrd m³ actually supplied, 37 percent was imported from the Netherlands, 17 percent from the Soviet Union, and 16 percent from Norway. Of the 52 mrd m³ consumed, the residential and commercial sector accounted for 30.3 percent, industry for 30.4 percent, electricity generation for 29.7 percent, and the remainder—chemical feedstocks, losses and own use, and so on—for 9.6 percent. Apart from the depressing effect upon natural gas demand of the economic climate in West Germany in the early 1980s, consumer reactions to higher prices and the need for energy conservation and use efficiency have also played their part, while the government has pursued policies favoring the use of coal and electricity from nuclear power plants. It seems that natural gas consumption will therefore be demand controlled in the period up to the mid-1980s, so that a level of not more than 57 mrd m³ is expected to be reached. In the longer term, into the 1990s, the federal energy program envisages natural-gas demand rising to between 60 and 64 mrd m³, of which the residential and commercial sector is expected to account for 35 to 40 percent, industry also 35 to 40 percent, electricity generation 10 to 15 percent, and the remainder 5 to 10 percent, so that emphasis will be upon the premium sectors, at the expense of under-boiler uses.

Looking now at the West German supply picture, although additional indigenous production could possibly materialize, production from existing projects is expected to decline to around 10 mrd m³ by the end of the century, by which time also imports from the Netherlands will have ceased altogether. Under import contracts existing in mid-1981, supplies from Norway will have ceased, and Soviet imports of 8 to 10 mrd m³ will still be available. It is probable that additional Norwegian gas will be available and that LNG from Nigeria may start being imported while, under the recently concluded contract with the Soviet Union, a further 10.5 mrd m³ will also become available. In the year 2000, therefore, taking both existing secure supplies and probable additional supplies, the quantity available, as things stood in early 1982, would probably be in the region of 40 to 44 mrd m^3 so that, even including the controversial new Soviet deliveries, there could be a shortfall of some 20 mrd m^3. Additional supplies are therefore going to be needed by West Germany during the next two decades, and should, for example, the delivery of any of the existing or planned volumes of Soviet gas not materialize, the share of natural gas in the overall fuels/energy balance would have to fall unless substantial volumes from elsewhere could be secured at an acceptable price. In West Germany, as in some other European countries, a considerable quantity of natural gas is sold in power generation and industrial markets where it has to compete with heavy fuel oil and with coal, and both of these tend to be priced considerably below crude oil. It would be unrealistic to imagine that natural gas priced at near crude oil parity would be able to find an acceptable outlet in these parts of the West German market, particularly with a trend in the early 1980s toward a buyer's market. This consideration would appear to indicate, for example, a reduced likelihood that Algeria might become a major contributor to the West German supply picture, at least so long as it maintained a policy of seeking crude oil parity, whether on a c.i.f. or f.o.b. basis, for its natural gas exports.

France, with proven reserves of well under 100 mrd m^3 and Italy with over 140 mrd m^3 have current production of 7 and 13 mrd m^3 respectively, but demand in both countries is substantially in excess of indigenous supply. In 1981, France consumed about 26 mrd m^3 annually and seems likely to require between 35 and 40 mrd by the end of the century, so that indigenous production's share of overall demand seems likely to decline from its current 30 percent to only 4 or 5 percent by the year 2000. Imports from the Netherlands were 8.8 mrd m^3 in 1980, but these will cease entirely shortly after 1995, and contracted imports from Norway of 2.7 mrd m^3 in 1980 will also have declined to below 1 mrd m^3 by this time. The Soviet Union supplied 4 mrd m^3 in 1980 and, with the recently concluded additional volumes due to come on stream in 1984/85, will be contributing around one-third of France's total supplies at that time. The other main source of imports for France is Algeria, which supplied just over 2 mrd m^3 in 1980 and is contracted to supply up to 9.6 mrd m^3 annually by 1990, or over 27 percent of expected

French requirements at that time. Nigeria may be providing up to 2 mrd m^3 by the early 1990s, but even if all the Algerian, Soviet, Norwegian, and Nigerian supplies currently expected do materialize, France seems likely to face a shortfall of up to 5 mrd m^3 of supply by the end of this decade which will increase during the 1990s if additional imports are not secured.

A similar situation faces Italy although probably not until toward the mid-1990s. The availability of Algerian imports is even more important for Italy's overall supply than it is for France. Indigenous production accounted for around half of Italy's requirements in 1980, but demand is likely to increase from 27 mrd m^3 to nearer 40 mrd m^3 by 1990 and to remain at this level through to 2000, and if indigenous production can be held at 10 mrd m^3 it will by then meet only 25 percent of demand. Algeria did not supply any gas to Italy in 1980, but, under existing contracts, pipeline supplies are due to build up to 13.3 mrd m^3 by 1990 and to stay at that level until at least 2000, and during most of this period they should represent about one-third of total supplies. In early 1982, deliveries of Algerian gas through the trans-Mediterranean pipeline were not taking place owing to a wide measure of disagreement between the two sides on pricing. The Algerians had decided to ask Italy for substantially higher prices than those originally agreed upon at the time of the signing of the contract for the construction of the pipeline. Italy also could not agree that Algeria should require for its pipeline deliveries the same high prices as it was insisting upon for its LNG exports elsewhere, and the two sides were at least $1.50/MMBtu apart in their negotiating positions in the spring of 1982. It seemed, by the autumn of 1982, that Italy and Algeria were on the verge of a compromise solution that would allow gas to start flowing through the hitherto unutilized pipeline. Italy received about 6.5 mrd m^3 from the Netherlands in 1980 but, as with all other recipients of Dutch gas, these deliveries will cease between 1995 and 2000. Libya supplied 2.4 mrd m^3 of LNG in 1980, but contracted volumes will decline by 1990 to 1.8 mrd m^3 and will cease by 1995 unless an extension of the contract is negotiated.

The other major supplier for Italy is the Soviet Union, with deliveries of 6 mrd m^3 in 1980 under an existing contract which could peak in 1990 at 7 mrd m^3 and decline to around 1 mrd m^3 by 1995. It is likely that Italy will become the third major buyer, after West Germany and France, of gas from the new Yamburg-Urengoi Soviet export scheme, having negotiated to purchase 8 mrd m^3 out of the projected 40 mrd due to reach Western Europe, although, in mid-1982, the Italian government was still withholding final signature, partly as an indication of its disapproval of Soviet action in connection with events in Poland. In the event that Italy and Algeria do reconcile their differences it is possible that contracted imports plus indigenous supplies could exceed envisaged demand for several years during the middle of this decade, and in this case Italy would perhaps find difficulty in accommodating the full 8 mrd m^3 due to be supplied from the Soviet Union. If, on the other hand, Algerian deliveries had not materialized, a shortfall would have occurred within a few years, and demand would not be satisfied even if the full 8 mrd m^3 of Soviet gas were to commence delivery in 1985. It is possible that small volumes of LNG from Nigeria might be available after 1990, but even if all contracted Algerian and Soviet volumes and those under negotiation do materialize, Italy faces a shortfall in the early 1990s that could exceed 10 mrd m^3 by the end of the century.

There are no other northwest European countries with known significant proven reserves of natural gas except for Denmark, where offshore deposits could amount to over 100 mrd m^3, but there is an existing demand for some 10 mrd m^3 in Belgium, 5 mrd m^3 in Austria, and 2 mrd m^3 in Spain. While this level of demand is not expected to change in Belgium or Austria, Spain envisages using some 10 mrd m^3 per annum in 1990 and 14 mrd m^3 by the end of the century, nearly 5 mrd of which should come from Algeria.

In 1990, Belgium will take about half its imports from Algeria and some from the Soviet Union, Norway, and Nigeria, as compared with the 75 percent from Holland and 25 percent from Norway that it took in 1980. It may be noted that Finland imports and consumes around 1 mrd m^3 of Soviet natural gas annually and that Sweden has had talks over recent years about the possibility of Soviet imports, although nothing has so far come of them. Ireland produces and consumes just over 0.8 mrd m^3 annually and although there are reserves of probably more than 100 mrd m^3 in Greece, production there, as in Spain, Portugal, Turkey, Switzerland, and Sweden, is negligible. Finally, although Austria produced some 1.4 mrd m^3 of its own gas in 1981, it relies for its imports of 3.2 mrd m^3 entirely upon the Soviet Union. See Table 5-2 for the latest figures for West European production, consumption, and imports/exports.

Table 5-2. European Production, Consumption, and Imports/Exports of Natural Gas, 1981 (mrd m^3 at 9,500 Kcal/m^3).

	Net Production[a]	*Gross Consumption*[b]	*Exports*	*Imports*
Austria	1.41	4.60	–	3.19
Belgium/ Luxemburg	–	10.08	–	10.08
Denmark	–	–	–	–
Finland	–	0.80	–	0.80
France	6.19	25.82	–	19.63
West Germany	16.62	48.90	1.95	34.23
Greece	–	–	–	–
Ireland	0.80	0.80	–	–
Italy	12.91	25.64	–	12.73
Netherlands	71.36	33.69	39.97	2.30
Norway	26.97	–	26.97	–
Portugal	–	–	–	–
Spain	–	2.30	–	2.30
Sweden	–	–	–	–
Switzerland	–	1.15	–	1.15
Turkey	–	–	–	–
UK	36.21	47.42	–	11.21
	172.46	201.10	68.89	97.62

a. The amount of gas collected from underground sources after the removal of sulphur and nitrogen and useful liquid fraction, with the deduction of quantities reinjected, where relevant.

b. Gross consumption equals: net production + imports – exports – stock change. Equivalent to total national consumption.

It will be clear by now that although some countries had their own natural gas industries based on indigenous production for nearly two decades, the major growth in continental West European consumption occurred as a result of the availability, at rather competitive prices, of gas from the Netherlands. It is perhaps ironic that the major supply problems faced by the same gas industries during the next two decades will be caused by the gradual decline and ultimate cessation of that very same availability of Dutch gas. A substantial investment in infrastructure has been made in West European gas-consuming countries, and since that infrastructure is capable of giving good service for many years to come and since, particularly in the domestic or public distribution sectors, it would be very difficult if not impossible to replace that infrastructure with one appropriate to another energy form, with the possible exception of electricity, it is clear that natural gas is going to be used in Western Europe for the foreseeable future in at least the volumes at which it is used today. The share of the overall primary energy supply provided by natural gas may only increase slightly from its present level of between 14 and 15 percent but, assuming even a very modest economic growth for the area as a whole, the volumes required will steadily increase as overall energy demand increases, even though the relationship between economic growth and energy consumption may change significantly as a result of conservation efforts. The decline of not only Dutch natural gas exports but also of indigenous production in all countries except Norway can only mean a significant increase in import requirements. It is over the capabilities and reliabilities of the countries from which these imports may be possible that so much discussion and controversy is currently raging, and it is to a more detailed examination of some of these countries that we now turn.

6 THE NATURAL GAS EXPORTING COUNTRIES

The geographical location of Western Europe makes it especially well situated to receive supplies of gas by both LNG tankers and pipeline from many of the significant existing or potential natural gas exporters, a few of which, such as Norway and the Soviet Union, have been considered elsewhere in this book. It is to a closer look at the other countries that are either already exporting or preparing to export some of their natural gas resources to Western Europe and elsewhere that this chapter will be devoted. Particular attention is paid to Canada, in view of its considerable potential as an energy exporter and its membership in the Western Alliance.

PIPELINE POSSIBILITIES

If one first considers the possibilities for pipeline deliveries it is apparent that, given the experience of long-distance operations in North America and the Soviet Union, even countries as far away as Iran and Nigeria could be considered viable contenders for the West European market. Equally, less distant countries in the Arabian Gulf such as Saudi Arabia, Qatar, and Kuwait may also be regarded, in terms of geographical proximity alone, as distinct possibilities. However, in the spring of 1982 not only was the pipeline linking Algeria with Italy not delivering the 13 mrd m^3 annually for which it was designed and built, but the second IGAT (Iranian gas trunkline) pipeline whereby Iranian gas from fields in the southwest of the country was to have been piped to the Soviet Union, and a similar quantity delivered thence to Western Europe, was not even constructed. While other projects were under some discussion—such as a further Trans-Mediterranean pipeline, this time between Algeria and Spain, and even a Trans-Saharan line from Nigeria and another from the Middle East via North Africa or Turkey—it certainly looked unlikely that any significant additional volumes of pipeline gas would reach the West European market from any of them, except Algeria, until well into the 1990s at the earliest and probably not before the next century. It may be somewhat puzzling that while pipeline gas from the Soviet Union has been flowing, almost without a hitch, from one side of the european ideological divide to the other for over a decade and in steadily increasing quantities, it had not proved possible by the end of 1981 to inaugurate any other pipeline supplies from outside Western Europe. What seems to be important is not so much the ideological or political color of the producing country, which may often find it expedient to put certain economic considerations before those relating to official dogma, but the control or lack of control over both domestic affairs and international relations of the regimes from or across whose country or countries the pipeline must pass. This factor alone enormously complicates the business of projecting, let alone constructing and operating, an international pipeline. Whereas it should normally be possible to negotiate with a given regime or regimes commercial terms satisfying both producer and transit-country interests, it is a very different matter when it comes to assessing the likelihood of those regimes adhering or being able or, indeed, allowed to adhere to those terms throughout the at least twenty-five-year lifetime of the project in the face of political pressures of one sort or another.

Clearly, the greater the number of countries between the producer and the consumer, the more difficult the project negotiation becomes and the greater the chance that some political instability will afflict some or other part of the venture. In the case of the Soviet pipelines to Western Europe, although it is possible that political unrest in Eastern Europe could increase to the point where mechanical interruption of supply could occur, the ability of the Soviet Union to control matters generally is clearly considerable. If, on the other hand, one looks at possibilities for Middle Eastern pipelines, other than those involving the Soviet Union as a transit/exchange partner, it is readily apparent that not only is one or other form of instability virtually endemic to the area but that there are a number of political, religious, ideological, or factional "divides" en route to Western Europe. A pipeline from Iran through Turkey to Greece and Italy might be considered a less trouble-prone route than some, but it would have to pass through the highly unstable Kurdish area to the north of Iraq, an area that does not seem likely to be pacified in the near future and one in which, even in times of relative peace between Iran and Iraq, there is plenty of potential for any pipeline to be used for political leverage. The construction of such a pipeline through the mountainous and somewhat lawless regions just south of the Soviet border would be a formidable undertaking in any case, given the poor access and difficulty of communications that would be entailed. However, considerable economic benefit could be brought to all the peoples involved, and it is therefore not out of the question that they might be persuaded both that it was in their interests to ensure the uninterrupted operation of the pipeline through their particular sectors and that they themselves should therefore provide an effective degree of policing—something that would otherwise be almost impossible to secure from outside.

Pipelines from the countries on the southern shores of the Arabian Gulf, if they were to avoid the densest concentration of instability represented by the interrelationships among Syria, Iraq, Lebanon, Israel, Jordan, and the PLO, would have to pass across Saudi Arabia into Egypt. Although it is possible to contemplate an improvement in relations between Egypt and some of the other Arab nations, in spite of or perhaps because of continuing progress on the Camp David Agreement, such a welcome change is less to be expected in relations between Egypt and Libya, the next country across which the pipeline would have to pass. The unpredictability of the Libyan leader, Colonel Qaddafi, is indeed one of the key factors militating against any such venture as a pipeline from the Middle East across North Africa to Western Europe.

Whether or not Libya's relations with all its neighbors would improve significantly under different leadership is a matter for speculation, but its militancy in anti-Israeli, proterrorist, and prorevolutionary activities from the Western Sahara and Morocco through Chad and the Sudan across to the Horn of Africa and up into Palestine, and its military links with the Soviet Union, have not been reassuring to potential investors in large North African projects throughout the 1970s and the 1980s. Even Tunisia, the last country through which a North African pipeline would have to pass before crossing the Mediterranean to Sicily and Italy, has experienced Libyan adventurism in the recent past. If, however, the gas-producing Arab countries could combine their interests to the point where they could exert sufficient influence upon the Libyan leadership to discontinue its boat-rocking activities, the prospects for a Middle East-Western Europe gas pipeline would be considerably enhanced. They might also be enhanced if, for part of the journey, the gas were converted to liquefied natural gas (LNG) in say, Egypt, and shipped to French or Italian terminals from the eastern Mediterranean, thereby bypassing that part of the journey across Libyan territory. Libya appears to have so damaged its political and diplomatic relations with its immediate neighbors and many countries on different continents that it now has few friends,[1] and even its close military relationships with the Soviet Union may no longer mean much to it if it has also lost political credibility among revolutionary circles. It is true that there are enormous quantities of Soviet arms stationed on Libyan soil, but a lot of these are reported to be suffering from neglect or misuse, and in view of the Soviet Union's own economic and financial problems, it is not now by any means certain that Libya will continue to receive unlimited Soviet military backing. Not only may the financial burden to the Soviet Union become too great, but Qaddafi's own unpredictability may discourage those in the Soviet leadership who might have cherished the notion of using Libya as a geopolitical instrument to further Soviet designs upon the African continent.

Animosity between Libyans and Egyptians appeared to have survived the passing of President Sadat, and if Egypt does not appear to be pursuing policies considered acceptable by Libya, there may be further Libyan attempts to destabilize the new regime there. The discovery of commercial deposits of hydrocarbons in the Western part of Egypt may provide a further focus for Libyan attention although not because of any envisaged shortfalls in Libya's own supplies. Libya exported about 3.6 mrd m^3 of LNG to Spain and Italy in

1. Roberto Aliboni, *Libya's Inter-Arab and Regional Politics* (Rome: Instituto Affari Internatzionali, 1981).

1980, but its official attitudes, as expressed in OPEC, against the production and export of more of its hydrocarbon resources than necessary to provide for its rather unpretentious economic development plans, it being a low-spending country, indicate that it would not be realistic to assume that production of its, mainly associated, natural gas would rise to even 10 mrd m^3 in the foreseeable future. As long as Libya behaves as it has been behaving in recent years, its involvement in any international economic venture of the political and strategic nature of a natural gas pipeline is hardly likely to be solicited or welcomed.

The problem posed by the geographical location of an unpredictable Libya may not actually have to be tackled by pipeline planners for some time, for by early 1982, forecasts of lowered energy demands generally and of reduced estimates of requirements for natural gas in Western Europe over the next ten or fifteen years appeared to have reduced active interest in any pipeline supplies from the Middle East that would be deliverable before the mid-1990s at the earliest. However, in spite of all the geopolitical complications briefly discussed in this chapter, it would seem logical that some of the one-third of global reserves of natural gas located in the Middle East should find an outlet in the West European market, and although it will probably do so, initially, in the form of LNG from Qatar, pipeline deliveries always remain a distinct possibility, particularly in a more stable era.

As mentioned earlier, consideration has been given to the possibility of constructing a pipeline from Nigeria northward through Niger and Mali and across the Sahara to Algeria to link up with other trans-Mediterranean pipelines so that Nigerian gas could flow to Western Europe by two routes, as is the case with Algerian gas. Because there has been no specific search for gas in Nigeria there is great uncertainty as to how much gas is likely to be available although known proven reserves are estimated to be in the order of 2 trillion m^3, mainly located in the Central Delta. The available reserves tend to be split over a large number of medium- and small-sized fields, and their development requires the construction of extensive gas-gathering systems, which involves considerable investment. About 1.4 mrd m^3 was consumed in 1980, and large quantities of associated gas were flared. It is planned to use a lot more gas domestically for electric power generation and industrial projects, while population growth seems likely to be in the region of 3 percent per annum and economic growth rates are expected to approach 5 percent. Since domestic use of gas is also intended to back out or replace fuel oil—so as to free oil for export—it is likely that demand will increase rapidly over the next two decades, and it has been estimated that the reserves necessary to meet such a demand would be in the order of 1.1 trillion m^3. This still leaves a substantial availability for export, with the prospect of a lot more being found, and it should be borne in mind that the flaring of associated gas after 1983 is prohibited. In 1981, something like 20 mrd m^3 of gas produced with the crude oil was flared, and alternative uses for this gas will have to be found if oil production levels are to be maintained. Export availability could therefore be some 30 to 40 mrd m^3 per annum in the 1990s, although in 1982 the Bonny LNG project was the only one planned for the utilization of both associated and non-associated gas for export, and it would have had a plateau production level of 21 mrd m^3. But work on the project was stalled during 1981 owing to general financial problems and although the long-term benefits to the country to be derived from this project would be considerable, there were formidable difficulties in providing the substantial resources required in the short term to get the project off the ground. These difficulties subsequently resulted in the liquidation of the venture.

The situation was not improved by the loss of revenue experienced by the government in early 1982 as a result of falling oil prices and reduced export volumes. At the time of writing, it did not seem likely that deliveries of Nigerian LNG would be made either to Western Europe or to the United States before the late 1980s/early 1990s, and any interest in a pipeline scheme had been temporarily reduced to zero. A trans-Sahara pipeline would not in any way be an alternative to the Bonny LNG scheme and would in any case require considerably greater investment. It might become viable as a second major export project once the LNG scheme was in operation, and it would be easier to find the finance for it if revenues from the Bonny project were being generated. The reserve base would certainly be there even allowing for domestic requirements and for the Bonny scheme, but a pipeline project would have to be evaluated against additional LNG export projects in any case, and it does not seem as though pipeline gas from Nigeria will reach Western Europe until well into the 1990s.

It must also be borne in mind that there might be some considerable geopolitical difficulties involved in such a scheme even if the technical, commercial, and financial problems could be overcome. A pipeline from Nigeria to Western Europe would cross the troubled Sahara, where the long-running West Saharan dispute involves Morocco, supported by the United States, on one side and the Polisario guerrillas, supported by Algeria and Libya, on the other. This dispute was, in early 1982, threatening the coherence of the Organization of African Unity, in which Nigeria plays a prominent role, and it is difficult to see how conditions stable enough to allow for the construction of a pipeline could be expected to materialize so long as the dispute remains unsettled. Relations between Libya and Nigeria have not been harmonious in recent years, particularly following Libya's invasion of Chad, and there are, apart from the basic problems associated with Arab relations with black Africa, a number of religious and tribal factional problems within Nigeria itself and elsewhere along the likely route of a pipeline. As a final consideration, Nigerian pipeline gas might threaten to back out Algerian gas in competition for the West European market, and it could be expected that Algeria, which has strong views on the pricing of internationally traded gas, might wish to have a say in both the timing and pricing of any gas transiting its territory.

North Africa does not seem, in the early 1980s, a particularly attractive part of the world, geopolitically speaking, across which to plan the construction of international pipelines carrying such a strategically sensitive commodity as natural gas, and it is difficult to see the situation changing for some time to come. Control over events there is not exercised by any single country, although there are one or two local countries that might have aspirations in that direction, and yet there are some extremely weak and poor countries that are highly susceptible to the influence of others and only too manipulable by those who have an interest in fostering instability for geopolitical reasons. Minerals, of interest to both the developed countries of Western Europe and certain developing countries, exist in significant quantities in some of the countries across which pipelines might be expected to pass, a notable example being uranium ore. There is a general shortage of water in the region and, further south, many years of insufficient rainfall have created refugee problems and food shortages, any of which could provide the spark that could destabilize large areas. The policing, let alone the construction, of a several-thousand-mile pipeline under such conditions would be extremely difficult.

Reference has already briefly been made to pipeline developments in the North Sea and it is quite likely that, over the remainder of this century, the network of pipelines there will evolve to the point where not only will Norwegian gas reach the continent of Europe by a number of different routes, for example to West Germany, Holland, and Denmark via Sweden, but the reserves in the U.K. sector will be linked with those in the Norwegian sector and with the Continent. This will result in a considerable strengthening of the security aspects of the West European supply network, although a great deal of hard political bargaining will be necessary before this desirable situation can be reached. However, in view of both party-political and national economic interests, it is difficult to escape the conclusions that the prospects for any significant additional European pipeline supplies of natural gas reaching the main West European markets before the mid- to late 1990s seem small.

Elsewhere, there are geographical and climatic problems yet to be overcome before pipelines can be built. There are numerous political difficulties, and in some cases the snags are financial or infrastructural. The element of uncertainty is in all cases rather high and probably sufficient to divert many potential risk takers into less uncertain energy projects even though many transit countries would stand to benefit economically. So Soviet and, to a lesser extent, Algerian pipeline gas will continue to dominate the West European gas import scene at least until the early 1990s, and it is probably not unreasonable to suggest that, in view of their inherent vulnerability, long-distance pipelines from the Middle East and West Africa are more likely to experience their heydays in the twenty-first than in the twentieth century. From the point of view of both the gas importer and the exporting country, for both economic and political reasons, the need to select the least complicated route is paramount, and the increasing lack of any clear-cut control over global or even continental events by any nation or group of nations means that every additional country that has to be involved in the international flow of natural gas complicates matters in a number of different ways. For this reason, in spite of the need for very much closer bilateral relations between the participating countries, LNG schemes appeared at the beginning of the 1980s to have certain advantages over pipeline schemes when considering future natural gas supply prospects for Western Europe.

GAS BY SEA

Critics of LNG schemes have tended to harp on the example of Algerian-U.S. developments in an attempt to demonstrate their fundamental unreliability, but there is no form of internationally traded energy that can be considered immune from security-of-supply risks and, short of piracy on the high seas or a deliberate act of war, in the final analysis LNG schemes only have to depend upon the maintenance of equitable relations between the minimum number of independent countries. At the beginning of the 1980s, Algeria had the potential to be involved in active LNG supply schemes with the United States, the United Kingdom, Spain, and France, although actual deliveries amounted to no more than about 4 mrd m^3 in 1980 in view of the disputes surrounding Algeria's policy of seeking post-contract price increases. Algeria claimed that the price then being paid for internationally traded gas offered little attraction to producer countries and that Algeria itself was netting, at 1979 prices, only about $1.25 per MMBtu at the wellhead, or one-third of the value in terms of thermal equivalence it derived from its high-priced, low-sulphur crude oil exports. The Algerians, therefore, not only promoted the idea that LNG should be sold on terms comparable to the cost of what they claimed was its real global economic alternative —namely, synthetic gas from coal—but sought to maximize pipeline exports for which they would endeavor, particularly in the case of Italy, to secure the same delivered prices as for LNG, that is, in excess of or close to $6.00 per MMBtu. This they were able to achieve in the case of LNG deliveries to Belgium and, after many months of stalemate, with France, but in so doing the Algerians delivered a blow to buyers' confidence in the security of LNG schemes and harmed not only their own reputation as a supplier country but also their export earnings. They may have felt more confident in their ability to pursue the course of action that they chose because of their assessment, in the late 1970s, that Western Europe would face considerable natural gas supply deficits by 1990 and that those would increase rapidly thereafter, so that there would be a sellers' market for those producers lucky enough to be within reach. As things turned out, by the early 1980s Algeria was having problems disposing of LNG from its existing plant, and while there were prospects that deliveries to the United States would resume on an increased scale sometime during the decade, contracted supplies to the Netherlands and West Germany seemed unlikely to take place before 1990. Algeria's approach to its LNG export policy seemed to be determined more by its aspirations to play a leading role in the evolution of a global LNG pricing system—perhaps with the creation of an organization similar to OPEC—than by its purely national interests, and it was apparent by the early 1980s that there would be few producer countries in a position to consider an Algerian pricing initiative favorably for several years to come.

In 1981, Algeria had proven reserves in the region of 3 trillion m^3, the vast majority of which was in the form of unassociated gas. Domestic consumption was low, some 4.5 mrd m^3, but rising rapidly, while exports of natural gas had been occurring since 1964, when the first deliveries of LNG were made to the United Kingdom. Algeria's overall population was increasing at the rate of 3 percent per annum, but the urban population was growing at twice that rate, indicating that the demand for gas would also continue to grow, particularly since the domestic use of gas was to be officially stimulated in order to free more oil for export. The country's stated development policy is to reduce emphasis on capital-intensive industrial projects, a policy that would seem to militate against further LNG export schemes in which the liquefaction facilities have to be built by the Algerians. Furthermore, as indicated elsewhere, Algeria has stated its desire to equalize the return to the country on its hydrocarbon exports, a consideration that was no doubt also influential in the Algerian decision to suspend its contracts with European companies for the construction of a third gas liquefaction plant at Arzew and the abandoning of another LNG plant project at Isser. However, Algeria's natural gas reserve base is sufficient to accommodate both an expansion of domestic demand and further increases in export capacity, and it seems likely that the full utilization of existing and the development of new export facilities will be most important financially for the fulfillment of Algeria's development plans. It is of interest in this connection that between 1975 and 1980, despite the real-terms increase in energy prices, Algeria registered a current account deficit of over $11 billion and borrowed heavily on international capital markets. It is not expected that oil revenues will rise in real terms in the immediate future, and with an oil reserves-to-production ratio declining to twenty years, oil revenues seem unlikely to be sufficient to satisfy the country's growing demand for imports. In 1979, LNG exports contributed about $0.6 billion to export earnings, but deficits between import expenditures and overall earnings could amount, in the mid-1980s, to many times that figure.

In 1981, operational LNG capacity in Algeria was estimated at nearly 26 mrd m^3 per annum although actual production was very much less. Total firm plant capacity was approximately 35 mrd m^3 per annum, while operational and approved contracts for overall gas exports amounted to some 53 mrd m^3 per annum, some of which would be met by exports through the trans-Mediterranean pipeline scheme which had an initial capacity of between 7 and 8 mrd m^3 per annum. A further 10 to 20 mrd m^3 per annum pipeline project through Spain was under study by an Algerian-Spanish-French consortium—SEGAMO—and the prospect for enhanced pipeline exports from this country seemed apparently better than those for LNG since the size of the investment in Algeria would be less and the netback to Algeria greater. In view of its proximity to the West European markets, the technology would be easier, the implementation faster, and the gas wastage less, and much would therefore depend upon the restoration of confidence in Algeria's intentions to uphold its side of any supply contract.

Countries that might become LNG exporters to Western Europe before the end of this century in addition to Libya, Algeria, and Nigeria include Canada, Cameroon, Qatar, Abu Dhabi, and Trinidad, of which Qatar and Cameroon seemed, at the time of writing, to be the most likely to bring a scheme into operation before 1990. It was expected that up to 10 mrd m^3 of LNG from Qatar might reach Western Europe, primarily West Germany, by 1987/88, while up to 5 mrd m^3 might be available from Cameroon by about 1990. Reserves in Trinidad have been established as sufficient to support an LNG project with a capacity of around 6 mrd m^3 per annum, although whether Western Europe will prove as attractive a market as is the United States to the government of Trinidad is not yet clear. Much will depend upon U.S. regulatory moves on the question of domestic natural gas. A Trinidad project could probably commence supplies by the late 1980s although substantially increased deliveries over 6 mrd m^3 are not thought likely to materialize in view of serious possibilities of using a lot more gas domestically.

In the event that the United States does not turn out to be a promising market for LNG exporters in the Western Hemisphere, the question then arises of what is meant by "Western Europe" when considered in terms of an alternative market. The quantities of LNG likely to be available, for example, from Trinidad would not be likely to stimulate a great deal of buyer interest in Western Europe as a whole, and it would be necessary to establish a clear purchasing requirement on the part of one or perhaps two consumer countries. Indeed, in marked contrast to the degree of cooperation among West European countries engendered by the pipeline projects from the Soviet Union, no similar degree of cooperation seems likely in respect of LNG import schemes in view of the often rather different requirements of the individual countries concerned. Only if very large export quantities of LNG appear likely to become available from any one producer country is there a likelihood that more than one Western European country would cooperate over financial and administrative arrangements. Short of that, given the buyers' market that seemed, in 1982, likely to obtain for some considerable time, any more distant producer country wishing to sell its gas to Western Europe will first have to convince one particular importing country of the advantages of its own supply potential over others, and only then examine whether other importers could be interested sufficiently to expand whatever initial scheme had been agreed upon in principle.

CANADA

Canada possesses sizeable hydrocarbon resources and, in particular, large reserves of natural gas that should not only enable domestic consumption to be increased significantly above levels obtained in the early 1980s but should also allow for substantially increased exports of both pipeline gas and LNG. By the end of 1981, Canada's reserves of natural gas were estimated at over 2.5 trillion m^3, comfortably more than those of the Netherlands, Qatar, or Nigeria and approximately twice those so far discovered in Norway. Domestic consumption in 1980 was in the order of 52 mrd m^3, representing 22 percent of total energy consumption, and actual exports to the United States amounted to just over 22 mrd m^3. The total production of 74 mrd m^3 indicated a reserves-to-production ratio of over 30 to 1 and, on the face of it, Canada seemed very well placed to play a larger role in the international trade in natural gas than that represented by its 12 percent share of that trade in 1980. Canada was already actively considering exporting LNG to Japan, and a scheme was under discussion that might ultimately lead to LNG exports to Western Europe, but it was becoming apparent by early 1982 that all was not well with Canada's natural gas industry, and it is instructive to examine the situation in more detail as an excellent illustration of how political factors can outweigh economic ones when the disposal of a country's depletable resources is involved. The energy situation would be greatly improved if only political issues could be handled more positively by governments concerned with the unlocking of natural resources, but it is often the case that no clear definition of objectives ever takes place or, if it does, the stated objectives are, for primarily bipartisan reasons, seldom pursued long or hard enough.

In the case of Canada, two basic energy policies have been apparent in recent years representing the views of the Liberals on the one hand and the Conservatives on the other, although it has been the liberal policy that has predominated throughout the 1970s and that, though under considerable pressure for modification, profoundly changed the direction in which Canada's energy matters were moving as the 1980s progressed.[2] Furthermore, there are two main factors that dominate the energy politics of Canada, the energy demands of eastern Canada and the struggle between the western energy-producing provinces and the federal government for control over the disposition of Canada's energy resources. While the demarcation line between policies does not run along any geographical feature, it can be said that Liberal policy has tended to reflect more the interests of eastern Canada and Conservative policy more those of western Canada. It is in the western provinces that Canada's energy resources are concentrated. Alberta, which in 1980 was producing 85 percent of Canadian oil, produced, along with British Columbia, nearly all Canadian natural gas. However, because of their small population, the western provinces have less political than economic power since they have relatively low representation in the central government, and in spite of its enormous economic significance Alberta, which has strong allegiance to the Conservative party, had little say in the Liberal government's energy policies. These

2. Melvin A. Conant and Eugenie Maechling, *Political Aspects of a "New World" Energy Policy* (Washington, D.C.: Georgetown Center for Strategic and International Studies, 1980).

policies have tended to serve those provinces in the most densely populated and industrialized parts of Canada to the east of the Great Lakes, which also have the largest net energy consumption. There is an obvious geographical parallel here with the Soviet Union, where some 80 percent of hydrocarbon reserves so far discovered lie to the east of the Ural mountains whereas over three-quarters of all energy is consumed by the European part of the country some two thousand miles to the west. There is also an analogous political position since there is considerable competition between Soviet Autonomous Republics for a larger share of the national budget, and those politicians in western Siberia, which produces so much of the country's energy, are increasingly protesting the fact that their allocation of funds for infrastructural improvements is not matched by the availability of those goods that would help to make life more worth living and that are largely consumed by those in the more "civilized" areas to the west.

Returning to Canada, the eastern region, which has the largest energy demand but very little in the way of developed energy resources except hydropower and coal, also has the largest voice in government. Ontario and Quebec, the stronghold of the Liberal party, have therefore exerted great influence over energy policy. It may be pointed out that they depend heavily upon imported oil. As a result of the 1973 oil price shock, the Liberal government established energy targets for reducing both domestic demand for energy and oil imports. These long-term objectives, established during the mid- to late 1970s, involved the reduction of oil's overall share in energy supply, to be brought about partly by a gradual rise in the price of domestic oil and partly by keeping the price of natural gas artificially low in order to encourage conversion from oil to gas. Conservative energy policy, as manifested briefly in 1979, aimed at the raising of Canadian domestic oil prices toward the world market level so as to not only curtail domestic consumption but also increase reserves and stimulate domestic production. The Conservatives hoped to eliminate oil imports by 1990 by means of applying the revenues generated by higher oil prices to increasing the availability of Canadian oil. The trouble with this policy from a Liberal party point of view was that it would mean that the eastern provinces, which they envisaged as continuing to depend upon significant quantities of imported oil until the end of the century, would eventually have to pay the equivalent of world prices for oil that they would be receiving from western Canada. On the subject of natural gas, the Conservatives naturally favored exports to the United States in view of the revenues that they generated, but the Liberals had developed a long-range policy of halting gas exports and eventually transporting most surplus western gas to eastern Canada through an extension of the trans-Canada pipeline eastward from Montreal.

In spite of marked differences between the basic Liberal and Conservative energy policies, a certain degree of compromise between the federal and western provincial governments was arrived at during the second half of the 1970s, when the energy industries reached record levels of activity and when major discoveries in the Arctic and off the east coast indicated that oil self-sufficiency for Canada might be achievable by 1990. This sense of common purpose evaporated in the autumn of 1980 when the Liberal government returned to power and issued the National Energy Program (NEP), which in effect committed the Liberals to satisfying the energy needs of the oil-consuming eastern provinces through both price controls on oil and government involvement in the international and domestic oil markets. It aimed to provide energy security for Canadians at the minimum price necessary, and, in respect of natural gas, the overall objective was to increase availability generally so that Canadians would not have to use so much oil.

Domestic gas prices were to be kept lower than those for oil so that, by 1982, they would be equivalent to 60 percent of crude oil prices, a factor expected greatly to aid market penetration at the expense of fuel oil. The extension of the transmission system eastward to the maritime provinces was to be completed by 1983, and the Canadian distribution network was to be extended generally so that all areas of Canada should have access to natural gas where economically feasible. The NEP envisaged that there would be sufficient natural gas available for Canadians for the foreseeable future, even if a moderate increase in demand developed as expected, but, whatever happened, Canada's domestic requirements would come first and export requirements second. In this connection, if there were a surplus of gas over the period for which export licenses already existed for export to the United States, that gas would be exported. If, however, a deficit should occur, the export level would be reviewed. Although it was envisaged that surpluses would continue in both traditional and new or "frontier" producing areas, the amount of those surpluses would be tightly controlled by the NEP, and thus the quantity available for potential exports would be substantially lower than that which could physically be made available after satisfying domestic requirements.

Another aspect of the new NEP reflecting firmly held Liberal views was the provision made for the Canadianization of energy companies whereby it was intended that, through the purchase by the government of several large oil and gas companies, there would be 50 percent Canadian ownership of the energy industry by 1990. To encourage the greater participation of Canadians in the development of Canadian resources exploration and production, incentives would be introduced for existing companies that were 50 percent or more Canadian owned, and existing incentives would be reduced in cases where ownership was more than 50 percent foreign.

The NEP stipulated that Canadian gas prices would not be linked to world oil prices since this would inhibit substitution away from oil, but there would be incentives to increase both reserves and production and encouragement for consumers to use natural gas in lieu of oil. It was considered that the natural gas industry would continue to have substantial excess production capacity and that the main problem would be one of finding markets, although it would not make sense to go for exports at the expense of growth in domestic markets. While some energy exports were considered desirable, the Liberals intended to give priority to domestic energy security to end the paradox of dependency upon oil imports by one part of the country when there were energy surpluses in another. Exploration for natural gas in the Canada Lands in the north of the country were to be encouraged, even though the domestic market might not actually need the gas for many years, as part of a "safety net" approach. In general, the NEP did not plan on the export of even the most northerly and, from the point of view of the domestic market, the most inconveniently situated reserves of natural gas unless the broad social and economic benefits resulting from that export could be seen to justify it, and the federal government announced its intention of exhibiting an extremely demanding attitude when assessing export proposals and of taking both environmentalist and other interests of the local inhabitants very much into account. It was intended that gas export prices would equate to world prices for imported oil, although exceptions could be made, as with the U.S. market, if these were considered compatible with overall Canadian energy objectives.

The issue of the export of Canadian natural gas has two main elements, one of which is the Liberals' long-range objective to increase the consumption of natural gas in the eastern provinces, and the other of which is the financing of the Alaskan Highway gas pipeline that the United States and Canada agreed to build in 1977. While Alberta would be interested in exporting its gas to the United States, the federal government would prefer to see it kept for the Canadian markets in the east although consumption there will not increase significantly until the extension to the trans-Canada pipeline is completed, probably in the mid-1980s. In 1980, natural gas producers in the western provinces were producing some 20 mrd m^3 surplus to domestic requirements, and since they also had substantial quantities in underground storage it was apparent that they faced considerable cash-flow problems and needed to get rid of the surplus. They therefore proposed that they be allowed to build portions of the Alaska pipeline project in advance of the Alaska-Alberta section so that surplus Alberta gas could be exported to the United States until replaced by Alaskan gas toward the end of the decade. However, the pipeline companies maintained that in order to finance the construction of the relevant portions, they would need a substantial increase in gas exports, and the National Energy Board, which determines the volumes and prices of hydrocarbons moving outside individual provinces and those for export, approved only substantially smaller export volumes than had been asked for. This could be said to reflect a compromise between the desire on the part of the federal government to secure revenues with which to finance construction of the pipeline and provide both jobs and an outlet for the Canadian steel industry and the intention to keep Canadian gas in Canada.

The significance of the struggle for administrative power between the federal government and the western provinces for pipeline exports of natural gas to the United States is thus quite apparent, and prospects for increased exports will depend largely upon political factors. But the relevance of all this to prospects for LNG exports to other parts of the world has yet to be discussed. In this connection it is of interest to consider the implications of substantial natural gas discoveries off the east coast of Canada. On the one hand the discoveries might indicate that the eastern provinces could one day become independent of gas imports from the west—in which case the need for any extension to the trans-Canada pipeline would be reduced, and the need for Alberta to export its gas to the United States would intensify. On the other hand, the federal government might persevere with its plans to use western gas in the east and be more amenable to suggestions for increasing exports of LNG from some of the northern and eastern areas. There is little doubt that strong feelings exist in Canada, particularly in the Liberal party, on the question of Canada's economic dependence upon her powerful neighbor to the south, and while the Canadianization policy put forward under the NEP may in due course reduce the degree of control exercised by U.S. companies in determining the course of Canada's energy developments, it does not seem likely to eliminate the dependence upon U.S. markets for natural gas exports. It should be noted here that although licenses existed in 1980 for the export of 34 mrd m^3 of natural gas, only 70 percent of this was in fact exported, largely because of the short-term surplus of gas and an export price of $4.94 MMBtu in the United States. In 1981, licensed volumes increased to 45 mrd m^3 although exports remained near the 1981 level. It could be imagined therefore that a Liberal government, to the extent that it favored increased natural gas exports at all, would favor diversifying away from the U.S. market, and this might suggest some heightened interest in the creation of LNG export schemes. Regarding the development of domestic demand in Canada for natural gas, bearing in mind that exports would be actively discouraged if demand were ever likely to match or exceed supply, it seems likely that, as a result of the NEP measures, natural gas consumption may grow at a rate of some 5 percent per annum during the 1980s, compared with an estimated annual consumption increase for all forms of energy of 2.1 percent. Natural gas will therefore probably account for between 25 and 30 percent of total energy requirements by 1990, when domestic consumption may have reached 90 mrd m^3 per annum. In spite of the substantial reserves in the western provinces, it seemed possible in early 1982, that the federal government would not encourage the development of production surpluses greater than those experienced in the early 1980s, although changes in the general regulations regarding surpluses for export were being actively considered. It therefore looked as though it should be to the frontier areas of the far north and the eastern offshore that potential European importers of Canadian LNG need turn for encouragement.

Established reserves in the Mackenzie Delta, the Beaufort Sea, and the Arctic islands were estimated at some 400 mrd m^3 in 1979, although several times this amount may ultimately be discovered. This volume is certainly sufficient to justify a number of LNG schemes if a market could be found and if government policy on declaring surpluses were to be modified. The federal government would be likely to approve exports from the frontier areas in sufficient quantities to justify the transportation systems, particularly if these were to go through the main consuming areas of Canada. In spring of 1982 it appeared as though gas from the east coast offshore could be available for marketing by the late 1980s, from the Mackenzie Delta-Beaufort Sea areas by the early 1990s, and from an expanded Arctic LNG project by the mid-1990s. There were, however, many technical, environmental, fiscal, and jurisdictional problems to be overcome, and large, high-risk investments would be involved. It may be recalled that, in 1982, there were three major natural gas movement projects involving Canada, but only one of these related to LNG. They were: the Alaskan system, to move gas from the North Slope of Alaska (Prudhoe Bay) via pipelines to markets in California and the midwestern United States and with the addition of a lateral in the Yukon, to move gas from the Mackenzie Delta-Beaufort Sea to the same area; The Arctic Pilot Project, to evaluate the feasibility of a surface transportation system for the purpose of developing Arctic islands hydrocarbons and minerals, under which natural gas would be transported by LNG tanker to the Canadian east coast provinces and a complementary volume of western Canadian gas would be made available to the United States; and The Polar Gas Project, to deliver pipeline gas from the Arctic islands and the Mackenzie Delta-Beaufort Sea to markets in southern Canada and the United States. In addition to the Arctic Pilot LNG project, which is primarily directed toward the U.S. markets, although it has potential for deliveries to Western Europe, three projects had been proposed to deliver LNG from western Canada to Japan. However, the tonnages involved in those likely to be given approval by the government were small, not exceeding a combined 4.2 million tons a year. It did not seem likely, in early 1982, that the Arctic Pilot Project would start delivering LNG for the Canadian-U.S. market before 1985 at the earliest, and it would clearly not be until the late 1980s or early 1990s that deliveries to Western Europe might become feasible. However, by 1982 it was already clear that projected conditions in the West European market for natural gas during the 1990s would be rather different from those envisaged when the scheme was originally discussed during the late 1970s. Even so, the possibility exists for a West European extension of Arctic Pilot, provided always that sufficient surplus gas is declared surplus by the National Energy Board.

Much of this chapter has been devoted to an examination of Canada's potential as an LNG exporter for the West European market because, perhaps even more than in the case of Algeria, the analysis reveals a number of extremely important aspects of the LNG trade in particular and the international trading of natural gas generally. First, the size of known natural gas reserves in relation to current or expected domestic consumption levels does not always provide a reliable indication of whether or not gas will be exported from a country. Second, particularly in Western democracies, the complexion of the party in power may profoundly influence the commercial environment in which the inherently long-term oriented international gas business has to operate. Third, uncertainty about the sort of changes that a government, however "friendly" or in the Western camp it may be, may or may not introduce in respect of the future development of a nation's energy resources not only makes planning extremely difficult for those organizations already involved in mobilizing those resources, but it also discourages investors from becoming involved at all. Fourth, it demonstrates how policy decisions in respect of one energy form, particularly oil, can influence the future development of another. Fifth, it shows how relations between two countries can affect the prospects for the development of a gas export scheme between one of them and a third country. Sixth, it shows how, as a result of the inevitably lengthy gestation period for LNG schemes, changes in market conditions over that period can significantly affect the attractiveness of a particular export scheme. Finally, it emphasizes the need for and the difficulties associated with setting up policy guidelines for local use commitments before granting export permits.

In the spring of 1982, the prospects for substantial new supplies of natural gas from new sources being available for West European markets, either via pipeline or in the form of LNG, before the end of the 1980s at the earliest, seemed poor. Not only were West European demand estimates being scaled down, but repercussions from the global energy price shocks of the 1970s were still being felt in a number of potential supplier countries, and the general air of uncertainty was not one in which investment decisions on such long-lived projects as natural gas schemes were being taken with alacrity. Prospects for the 1990s seemed rather better, at least from the point of view of West European demand, and it appeared as though domestic problems in a number of potential producer countries might have been resolved sufficiently to allow progress on natural gas export schemes to be made. However, the international situation was rife with unresolved situations, and while these seemed more likely adversely to affect potential international pipeline schemes, they also cast shadows over the LNG scene. It is with an examination of the state of "the International Order" at the beginning of the 1980s that the next chapter is concerned.

7 THE CHANGING INTERNATIONAL ENVIRONMENT

It has by now become standard practice in the art of deflating futurologists to ask them whether they were among the 1 percent or fewer who accurately predicted the overthrow of the Shah of Iran. Fewer people, probably, would have thought it likely that Britain and Argentina would have gone to war in 1982 over the Falkland Islands and that the British foreign secretary would have resigned because someone, somewhere misread the entrails. Only a handful of visionaries could, at the end of World War II, have believed or even considered that within a quarter of a century men would be walking on the face of the moon or that smallpox would have been virtually eliminated from the face of the earth. Man's ability to surprise himself does not diminish, and it is often rewarding to ask a wise man what would not surprise him over the next twenty years. It is, indeed, a sign of wisdom today when a person declines to make forecasts about a wide range of issues—particularly those in the field of international affairs.

Less than a hundred years ago it was possible, at least if one belonged to one of the small group of world powers who basically were the global system, to take, with some confidence, a hundred-year view of how global events might develop. Fifty years ago, following World War I and the Great Depression, this was no longer the case, and today there are problems enough in trying to identify what the global system is in which the events are due to take place, let alone attempting to say which ones will occur when or where. One aspect of all this is that the unfolding of events has acquired an immediacy today that it did not have yesterday. People's senses are now becoming so assailed by the multiple impacts of so many events that they have difficulty in sustaining in their minds the background or reference picture against which they must evaluate possible future developments. It is difficult at the best of times for a swimmer to stargaze when treading water, and when the wind gets up or the piranhas start to nibble, his or her attention turns to more immediate problems. It is hardly surprising today that the thoughts of politicians and statesmen are most often turned from the far distance, in space and time, to the here and now even in countries in which they have traditionally taken a long-term view. On a somewhat less exalted level, those concerned with, for example, making the future for natural gas come true, may well despair at the diminishing availability of reliable predictions about the supply and demand for various types of energy. It might be of assistance at this point to try to recognize some of the forces that seem to be shaping the environment and to attempt to identify some of the reasons why the glass through which we now see so darkly is also evidently taking on the properties of a distorting lens.

About the only thing of which one can be certain is that the environment is always changing and will continue to change, but what is happening in the world of today is that the rate of change is accelerating and the directions in which change can proceed are becoming more numerous and haphazard than has been possible in the past. It would help the task of looking at the effect of changes if one could have a clear picture of what things are like before the changes occur, but, in the case of international affairs, if one tries to investigate even the most recent past it soon becomes obvious that what one sees depends largely upon what one has been conditioned to see. It is all too easy to see what someone else has prepared for one to see, and the effectiveness of one's vision also depends upon how well placed one is to see something. A well-educated, reasonably well-off West European or American, having access to as much information or bombardment from the media as he or she can handle, will have a very different view about the "natural" order of things than, say, a stateless Palestinian, a black African in Johannesburg, or a Soviet bureaucrat, and there can be no universally accepted "correct" view of anything that so defies definition. Furthermore, the growth in the global population and the finite nature of global resources of food, water, minerals, and space are creating situations where, for increasing numbers of people living in poverty and hunger, changes in the established order can only be for the better and are thus to be encouraged, worked for, and even died for. For those who have been privileged to enjoy the highest standards of living and comparative freedom of action and thought, any changes are increasingly likely to be interpreted as being for the worse and to be resisted as long as possible, by force if necessary. It happens, moreover, that those with access to and control of the major methods and networks of global communications are mainly those who are, as a result of education and experience, in a position to perceive and deplore the growing threats to what they have been conditioned to think of as the natural order—an order that has served them well in the past but that seems less likely to do so in the future. The more far-sighted among them appreciate the inevitability of the change to their system and are confident that good can be made to come of it, but the majority is mainly concerned with resisting change that they cannot bring about themselves and preserving the status quo in any way they can. There are even a few who seem determined to try to re-create the status quo to which they have become accustomed but which probably vanished forever with the advent both of weapons of mass destruction and of the means of mass information and manipulation, particularly in hands other than their own. The enlightenment that the advanced nations strove so hard over the centuries to bring to the "less fortunate," through education and improved communications, also served to bring home to them the blatant global social inequalities that had existed for hundreds of years and that still exist. It also raised the aspirations of countless millions far beyond the capacity of existing systems to satisfy them, at least within their expected lifespans. The level of global dissatisfaction has therefore risen over the nineteenth and during the twentieth century, though few people appear to have both the power and the inclination to do very much about it.

Quite apart from their desire and ability to spread enlightenment and impose cultural integration on weaker nations, the world's great powers have experienced a virtual monopoly of the means of communication, and a high degree of control over the global trading and financial systems remains largely in their hands even today. The ability to exercise military power on a global basis to secure the needs of an empire required a carefully integrated logistical system and network of reliable bases. It also required both massive and growing financing and, thus, the acquisition of still greater wealth. All this necessitated the development of a global approach to international affairs. Throughout the nineteenth and well into the twentieth century, one European power predominated in the directing of the global system. It is probably fair to say that only with the flowering of the British Empire did a truly global approach to both economic and political issues develop, since such issues had hitherto been mainly relevant to continental or intercontinental relations. The first global order was therefore one that was instituted and held together by one West European power, and though disputed over the years by several other similar powers and resisted by a few non-European regional powers, it was essentially European in character and was imbued with a belief in the fundamental correctness of European values. European control of the international order had, however, come to an end by the middle of the present century as a result, largely, of two world wars fought mainly among competing European powers, and even though the British emerged on both occasions on the winning side, Britain was so weakened that the empire could no longer be held together. Its disintegration marked the beginning of a major transition period in global events.

A GOLDEN ERA?

In the twenty-five years immediately following World War II, there emerged both the United States, already an economic giant, and, more recently, the Soviet Union as global powers, each being capable of bringing to bear at any one time decisive military, political, and economic leverage in just about any part of the globe except in each other's prime spheres of influence. Until probably the early 1970s, the control or management of the international order was indisputably in the hands of the United States, and a war-torn Europe was only too happy, given the alarming flexing of national muscles on the other side of the Iron Curtain, that its transatlantic benefactor had the economic, military, and technological ability to "mind the global shop." The order that prevailed during the first few postwar decades was not appreciably different from that which had existed previously, and it resulted in an exceptional period of growth in international trade, technological and industrial advance, energy consumption, the development of sophisticated armaments, and the communications explosion, in which probably all nations in the world shared and from which all probably benefited. The free-market system still determined the majority of trade flows, and the global monetary system was still under the near-total control of West European and U.S. financial circles. Economic growth of the industrialized Western nations still depended upon the creation of markets elsewhere in which the surplus production to which the capitalist system, according to Marxist teaching, inevitably gives rise could be sold, although the markets concerned were becoming perhaps a little less captive than they used to be. The use of military force to settle disputes between nations was becoming somewhat tempered by a Pax Americana in which the role played by U.S. forces themselves was significant, particularly in the Far East and Indochina. But the great European powers steadily withdrew from nearly all their military bases around the world, gave independence to their former colonies whether or not they were considered ready for it, and concentrated on the economic recovery of Western Europe. They benefited not only from economic aid from the United States but also from the defense umbrella that the United States, by virtue of its then-overwhelming military superiority, was able to provide. The United States, probably reluctantly, found itself appointed as global policeman and chief defender of the capitalist faith, but it shouldered the responsibilities of leadership and enjoyed a few decades of almost unnaturally great political, economic, and military influence.

Many people in the Western World today look back upon the twenty-five years following World War II as something of a golden era during which even the developing countries had a chance to prosper, particularly with the elimination of many hitherto crippling diseases and the enhanced results of agricultural research and pest control. Many of these countries came to experience independence from colonialism and imperialism for the first time although some of them found themselves saddled, on the departure of their former masters, with unresolved territorial or ethnic/racial problems at a time when they were both militarily and economically too weak to stand on their own feet. They needed somewhere to turn for protection and aid that, hopefully, would not condemn them again to what some of them were learning to refer to as capitalist exploitation. The emergence, therefore, of the Soviet Union as a global power heralded the start of an era of geopolitical game playing on a hitherto unparalleled scale and created for the first time a basis for the global polarization of the affiliations of less powerful nations. The ideology of Marxism-Leninism held a powerful attraction for newly independent countries, anxious to find a new political star to follow, and the Soviet Union as a nation not only possessed a comparatively untarnished colonial record but it was militarily strong and economically rather self-sufficient. It was easy for the Soviet Union, in the pursuit of its long-term ideological and geopolitical objectives, to profess altruism and to pose as the natural friend of those newly fledged developing countries whose regimes were keen to convince themselves and their people that independence from former rulers need not mean dependence upon new ones.

POLARIZATION AND POWER-DRAIN

In the 1950s and 1960s an essentially bipolar geopolitical confrontation emerged, based upon a fundamentally different set of philosophical concepts about how human beings should be governed. Few countries were in a position to remain aloof from the polarization process, and the ideological incompatibility between the two sides has remained the basis for the ongoing East-West confrontation that threatens the very survival of any world order today. The management of the East-West confrontation at levels short of outright armed conflict has preoccupied leaders of the world's most powerful nations for several decades and has involved virtually all other nations, however reluctant they may have been to get involved. The bipolar nature of the confrontation and the extreme differences between the sides has unfortunately meant that the West has tended to construe any setback to its own position as having been caused by the Soviet Union and any advance by the Soviet Union as an indication of failure on the part of the Western Alliance. In fact, the coherence of the Western power bloc has always been bedeviled by a major and essentially capitalist shortcoming, namely, competition among its member nations in furtherance of national interests. It can be argued that many of the real or imagined setbacks for the West have been less a result of any active exercise of geopolitical power on the part of the Soviet Union and its largely unwilling allies than the result of mistaken or misapplied policies adopted by one or other Western country in pursuit of its own ends and often in disregard of the real interests of the group of nations to which it, voluntarily, belongs.

The emergence during the last three decades of so many new and independent nations, each with full sovereign rights and a vote in the United Nations and other global institutions, enormously complicated the business of maintaining some sort of control over the international order. These young nations were quick to appreciate the benefits to be derived from playing one superpower against the other, given the immense importance that the superpowers attached to preventing each other's gaining at either's expense, and the voices of the least powerful countries of the world were therefore often heard as loudly in global deliberations of international matters as those of the most powerful. It also became apparent during the 1960s and 1970s that the most powerful nations were becoming increasingly less able or willing to use all the power at their disposal. The United States, for example, even with the support of its allies, was apparently incapable of bringing that power to bear to solve international problems in a manner favorable to itself, and so was in danger of losing the political will necessary to ensure the cohesion of the Western Alliance. The United States was still endeavoring to control geopolitical events in the early 1970s, but it had by that time already demonstrated in Vietnam, over the Suez Canal crisis, and in association with the Soviet invasion of Hungary that it was not prepared to use sufficient resources to decide the outcome. Such action as it did take over these and other events gained it few friends and quite a number of enemies, particuarly among its own younger generations. It may one day be seen as one of the great strengths of the Western democratic system that so much open criticism of and opposition to U.S. government action both at home and abroad could be tolerated, but the degree of paralysis in decisionmaking on matters of international significance to both the United States and the rest of the Western Alliance, to which such criticism gave rise, did mean that the Soviet Union was increasingly free to flex its muscles and step up its geopolitical activities in parts of the world in which it hitherto had no part to play. With the exception of Khruschev's setback over the Cuban missile crisis and his failure to secure the lasting allegiance of mainland China, the Soviet Union managed slowly but surely to strengthen its influence offshore, while at the same time rendering its own borders and those of its satellites in Eastern Europe less vulnerable to political, military, or economic interference from the West.

By the end of the 1960s, the United States and most of its allies felt sufficiently chastened by what they considered to be numerous geopolitical setbacks and adverse changes in global sentiment to feel that perhaps the time had come for some degree of reconciliation with the Soviet bloc. The Soviet Union for its part was by this time feeling sufficiently strong, politically and militarily, to be able to reciprocate without fearing either economic or ideological repercussions. Soviet foreign policy had for some time been concentrating upon supporting those regimes that, for whatever reason, were no longer prepared or obliged quietly to follow the geopolitical lead of the West, and in certain cases—notably, Ghana and Egypt—it was temporarily able to secure the installation of regimes that were at least prepared to express a pro-Soviet and anti-Western attitude. However, it must already have begun to dawn upon some in the Kremlin that Marxism-Leninism was not catching on as the vastly superior alternative to Western imperialism/capitalism quite as pervasively or conclusively as they had been led to expect. The Soviet leadership had perceived, and earlier than did United States administrations of this period, that blind support of regimes that were enemies of your enemies did not necessarily mean that such regimes would automatically become your friends. The Soviet Union therefore developed a policy of support for what they termed "progressive" regimes which, even if they were not about to follow an orthodox Moscow line, were anxious to indulge their new-found independence by emphatically not adopting a capitalist approach either to geopolitical issues or to their own economic and political development. It was still too early for a full evaluation of the consequences of outright acceptance of Soviet bloc economic and political aid to be made by those third world countries that had experimented with it, although preliminary indications, even with the more notable examples such as Cuba, could not have been very encouraging to Moscow. Indeed, it was possible to discern the first signs of disenchantment among third world countries with Soviet advances, and it was coupled with a feeling that perhaps neither superpower had all the answers.

By the early 1970s, however, the essentially bipolar relationship between the two superpowers was evolving, with the growing economic and political power of Japan and China in the Far East and of Western Europe in its own right, toward a pentagonal structure in which it was obvious that the Soviet Union had no significant friend or ally now that China had not only declined to accept Soviet ideological leadership but had agreed to open the "Bamboo Curtain" and receive overtures of friendship from the Nixon administration. The geopolitical stage was thus set for a completely new set of relationships, the full implications of which, especially in the fields of international trade, energy security, and defense, are still not fully clear a decade later. One aspect that was clear, however, was that the Soviet Union perceived itself to be the object of an encirclement conspiracy between the former imperialists and China, and its leaders resolved to do all they could to prevent, interfere with, or destroy any relationships between the other world powers that might already exist or that might develop in such a way as to add to the intense traditional Soviet sense of isolation.

THE DECADE OF DETENTE

The decade of the 1970s was nevertheless marked by the inception of détente, its flowering, and its decline. It was also, incidentally, marked by major shocks to the global energy pricing and supply systems and by strains within both the Western Alliance and the group of Communist nations. However, détente can probably best be seen as a concept for managing the East-West confrontation at levels short of outright armed conflict. It was embarked upon at a time when one superpower was undergoing a severe moral shakeup following its disastrous involvement in Vietnam and the traumatic effects of Watergate, and the other superpower was struggling with chronic problems stemming from the failure of its political system to cope with deteriorating economic situations both at home and among its allies. The United States, for its part, was experiencing declining geopolitical and geoeconomic capability, while the Soviet Union was beginning to savor increased political and military influence beyond the borders of its East European empire. The United States may have reasoned that it could bargain for some easing of the geopolitical pressures that it was facing with some economic quid pro quo, of which it was considered that the Soviet Union was in great need. The Soviet Union probably saw détente both as a means of reducing the risk of a nuclear confrontation with the United States and as a method of facilitating an accelerated West-to-East flow of technology, finance, and consumer goods that were needed to alleviate its economic problems. A superpower dialogue could moreover only enhance the Soviet Union's international standing, and Moscow never intended that détente should get in the way of Soviet long-term objectives to extend its hegemony whenever possible by the promotion of the ideological battle by any means short of war.

As the 1970s drew to a close it was clear that détente was moribund. The United States was disappointed to find that the geopolitical pressures were not abating and that some of its main European allies were apparently succumbing to Soviet political blandishments to the point where they appeared more concerned with the short-term economic benefits to be obtained as a result of improved trading relationships generally between East and West than they were with the long-term implications for the cohesion of the Western Alliance and of a materially and militarily strengthened Soviet Union. Soviet ideological fervor was not apparently diminishing and its military capability was being steadily and systematically enhanced, largely, it seemed, as a result of the increased trading opportunities afforded by détente. While the Soviet Union was clearly appreciating the expanded West-to-East flow of technology and capital, even though it probably never reached the levels expected of it by détente's most active proponents, the subscribers to a harder line in the Kremlin may have felt that, on balance, the Soviet Union was paying a political penalty for détente, in that it was of necessity inhibited from taking advantage of all the geopolitical opportunities available to it so long as it was obliged to maintain at least the appearance of a reformed character. Only in one or two West European countries did official support for détente still linger in the early 1980s, associated with West Germany's Ostpolitik perhaps, and with those left-wing elements who have never been particularly concerned about Soviet expansionism or the possibility that it could lead to the incorporation of Western Europe into the Soviet sphere of influence. As far as the United States was concerned, détente was dead even before the Soviet invasion of Afghanistan, and successive administrations were perplexed and disillusioned both by the ambivalence of the European allies over the East-West confrontation and by what they considered to be the two-faced behavior of the Soviet Union, which was seen as having benefited hugely from détente, both economically and geostrategically.

It will be recalled that, in the early days of détente, in 1972/73 a number of measures were discussed between the superpowers that were to serve to strengthen the improving relationship between them. One of these, the "economic underpinning of détente," was to be achieved partly through massive cooperation between East and West in the field of energy. At that time, a number of schemes were discussed involving the development, with Western financial and technological assistance, of some of the Soviet Union's enormous reserves of fossil fuels and the subsequent flow of oil, and possibly natural gas, from the Soviet hinterland to the "West"—Europe, Japan, and even North America. For largely domestic reasons on both sides, nothing came of the schemes although the turmoil into which the global energy situation in general was thrown by the events of the winter of 1973 obviously played its part in muddying the waters. In the United States, Watergate not only brought to an end the time in office of one of the chief protagonists of détente, President Nixon, but it set in motion a rethinking of the role of the United States in international affairs and clearly complicated its conduct of relations both with its allies and with the Soviet Union. In the Soviet Union, shortcomings in the performance of the oil-production industry in particular, and in other sectors of the economy in general, caused the Soviet leadership to rethink their energy policy. One outcome of their deliberations was a policy of developing Soviet onshore energy resources to the greatest extent possible without the large-scale involvement of outsiders, although the latter would be invited to play a role in the development of offshore resources and in the provision, under limited and carefully controlled commercial conditions, of those parts of the energy distribution infrastructure that the Soviet Union was clearly going to have difficulty in providing on its own.

It is of interest to recall that only about a decade ago the United States was able to contemplate, apparently with complete equanimity, substantial involvement in such a strategically significant field as the development of the Soviet Union's energy resources. Whether the Soviet Union really intended that this should ever happen in practice will probably never be known, but it is perhaps significant that the more recent manifestation of East-West energy cooperation, the Yamburg-Urengoi gas export project, never envisaged any substantial direct involvement of the United States either materially, except in the provision of certain technology, or financially. This may say something about the inherent difficulties of promoting economic cooperation between two superpowers of fundamentally opposing political and ideological views, important though it remains to attempt to overcome those difficulties.

EAST, WEST—NEITHER BEST?

This chapter was not intended to deal primarily with the evolving superpower relationship, but, to the extent that, in the eyes of many observers of the international environment, that relationship is likely to influence so many other areas of change for the foreseeable future, it is felt that certain aspects should be further illuminated as being essential parts of the background against which future changes, particularly in the energy environment, should be evaluated. It is a matter of concern on the part of some Europeans that so much time and effort has been and is still being devoted to the East-West geopolitical relationship when, it is argued, far more attention should be devoted to, for example, North-South problems, since these will ultimately be more significant in determining future global developments. Others who feel that the East-West confrontation has been overemphasized suggest that the future of global energy supplies should occupy the attentions of Western leaders more than it currently does. However, there is a powerful body of opinion—originally centered in the United States but, particularly since the Soviet invasion of Afghanistan, emerging elesewhere—which has become convinced that the Soviet Union is determined to use the relative energy resource independence that it enjoys in contrast to its ideological opponents to exacerbate the latters' geopolitical difficulties. It is argued by such people that the Soviet Union will do all it can do to destabilize the Middle East and any other area that might be of vital energy or resource interest for the Western Alliance.

As the world entered the 1980s, relations between the two superpowers could seldom have been more strained, with a new administration adopting a harder line than ever toward that nation from which, as in the time of the Cold War, most of the evils facing the United States were ultimately considered to have originated. The Soviet Union appeared in no way either to have changed its policy of promoting world communism by every means open to it or to have abandoned its objectives of attaining the status of supreme nation, however long that might take. It was, however, facing even more serious economic shortcomings, particularly in agricultural production, than at any time in recent years, and the unrest that had clearly existed in some of its East European allies for many years finally came into the open in Poland and threatened to do so also in Romania and possibly elsewhere, largely as a result of a growing perception of the failure of the economic policies pursued by the Communist regimes there and the frustration felt by the rest of the populations in view of their inability to do anything about the problems their countries faced under those regimes.

It would probably be too simplistic to assert, as some harder-liners have asserted, either that the Soviet Union has a master plan for world domination or that, faced with chronic economic and, more recently, social problems at home, it might feel obliged to create a major diversion by taking military action abroad. But it is difficult for many in the West to avoid the conclusion that the Soviet Union has been pursuing a highly directional opportunistic policy aimed at exploiting each and every geopolitical situation to the extent that it has been allowed to by the West or has been able to afford to in view of its deteriorating economic situation and the rising costs of its ongoing support for countries like Cuba, Ethiopia, and Vietnam. The Soviet Union has undoubtedly made geopolitical mistakes in the past that must have increased its awareness of the limitations of being a superpower, but it entered the 1980s very far from being a spent geopolitical force, and it must have also been well aware of the declining power of the United States and of the divisions within the Western Alliance–divisions that it had played a significant part in bringing about. Superpower competition for the political allegiance of countries in the third world and of the nonaligned bloc can be expected to continue to be major source of geopolitical issues throughout the 1980s, and each superpower will no doubt continue to try to take whatever steps it can to ensure that no political or strategic advantage will accrue to the other. This can only have a destabilizing affect in an already unstable international environment in which the control of events has passed beyond the reach of either superpower.

The diffusion of power away from the great European powers initially and the United States subsequently has resulted in a fundamentally changed international environment, and while it would be a mistake to believe that the Soviet Union has inherited an excessively large share of that power, the polycentric distribution of global power that now prevails makes it all the more imperative for those most dependent upon the smooth conduct of international relations, and particularly those most involved in international trade, to adapt to the new circumstances. It is understandable that nations should attempt to influence the directions of change in their own favors, but the management of international relations in a world in which it is increasingly difficult to determine what are desirable directions requires different approaches to those that appeared to work satisfactorily when a recognizable orientation was imposed upon the development of events by one nation or dominant group of nations. Preoccupation with the real, imagined, or threatened activities of the Soviet Union will probably prove an increasingly less productive approach—certainly less productive than efforts not only to put one's own house in order but also to establish clear national and alliance objectives.

It is encouraging that a number of formerly very powerful countries have come, often through bitter experience, to recognize their own limitations as actors upon the global stage. It is also encouraging that a number of newly industrialized countries have come to play a significant role in regional events. Indeed, the growing regionalization of international affairs appears to be one of the more noticeable changes in the global environment. It is possible to identify trends toward regionalization in most parts of the world—Latin America, Southern Africa, Southeast Asia, the northeast part of Asia, Eastern and Western Europe. In some cases, countries take a deliberate political decision to associate with neighbors, for largely economic reasons. Some examples of such alliances are The Association of Southeast Asian Nations (ASEAN), the Andean pact, and the European Economic Community (EEC). In others cases, one country, by virtue of its size, population, or economic or military power, becomes a focal point for activities in its region in a way that was not possible when the world was "managed" from London or Washington. Examples would be South Africa and Japan today and possibly Brazil and Nigeria tomorrow. The Middle East will, of course, continue to represent a most notable regional phenomenon although whether it would continue to do so were it not for its oil wealth is open to debate. Paradoxically, although there are so many disruptive factors at work in the region, this very fact gives it a particular identity unlike that of any other region in the world.

One implication of the trend toward regionalization is that not only will one or two countries in a region attempt to set up their own order, but they will tend to mobilize the support of their neighbors to prevent local strategic and security situations developing to the point where either superpower feels it must become involved. They will do all they can to keep the superpowers out, but should one superpower inflict itself upon the region, they will almost certainly attempt to involve the other in order to try to preserve a balance of impotence on the part of outsiders. Another implication is that, particularly in the area of trade, some of the older regions of the world, such as Western Europe and the Indian subcontinent, may find themselves by-passed, as the new regions attempt to "go it alone" or develop closer links with other new regions. The Pacific Basin may represent a quarter of the globe where this is already happening, with the focus of economic attention shifting from the North Atlantic westward, to where many of the newly industrialized countries and Japan are located.

Another important part of the environment that is changing in ways that can profoundly affect the business and political climate over the next two decades is that of human values. Societal research reveals that, prior to the recession in the early 1980s, in the industrialized countries of Western Europe and North America, people's attitudes toward government, big institutions, their work, and their play were all changing. While no two countries necessarily manifested the same trends, among the most notable were: a declining respect for or trust in the ability of governments to satisfy people's real requirements, a tendency for people to opt out from the bigger institutions and to form their own smaller networks of cooperation and communication; a reduction in the significance of the work ethic, and a requirement for more time for leisure and recreation. The repercussions from the recession may have further influence on the direction and rate of change, but change there will continue to be. Anyone who has lived in a former colonial power that has lost all its colonies, or in a great power that has not used its power and has as a result been humiliated, will be familiar with the feelings of guilt, shame, or frustration that can pervade a nation. All sorts of traditions and attitudes of the Establishment become questioned and criticized and sometimes condemned. Nations begin to wonder how they came to do what they did to other nations and become almost self-flagellatory in outlook. These national moods pass in time, but they can have fundamental repercussions on national and personal prestige and upon the quality of decisionmaking of the nations' leaders. This, in turn, can effect the way in which other nations believe they can deal with one, but it can also lead to a global reassessment of past performance and, above all, to a review of all those national values that were ultimately responsible for the past deeds and the manner in which they were carried out. Attempts by, for example, the West Europeans to impose the Protestant work ethic upon black Africans, or by the Soviet Union to impose its brand of communism upon the China of Mao Tse-Tung were largely unsuccessful and did not long endure the departure of the military presence or threat that accompanied imposition. But they are still being criticized today, as demonstrated by the ideological split between the Soviet Union and Communist China and by the strident and often apparently irrational manner in which the voices of former colonies and dependencies are raised in public chorus.

NORTH AND SOUTH—INDEPENDENCE AND INTERDEPENDENCE

Western values are by no means always desired by developing nations; nor are they always relevant to the situations that exist there, and it is possible that the religious fundamentalist revivalism that has characterized the revolution in Iran and that threatens the security of a number of regimes in the region will be paralleled in other parts of the world as the ability of those in power to persist with the imposition of alien social, political, or cultural doctrines upon unwilling and increasingly vocal indigenous populations is impaired by time or by the weakening of political will. Indeed, one of the major impediments to the "transfer of technology" so loudly but often misguidedly demanded by the politicians of some developing countries is the absence in those countries of appropriate infrastructural and human resources to absorb and benefit from, for example, equipment and processes developed in and for countries with totally different lifestyles, standards of living, levels of education, and approaches to work. It seems to be a common misconception that if the developed, industrialized countries of the West were, in partial expiation of some of their former colonialist and imperialist "sins," to hand over their technological secrets and unlimited funds to those whom they once exploited and otherwise mistreated, the latter would be able to "take off" economically, virtually overnight. Of course it is not actually within the gift of Western governments to make available their countries' technology, which resides in minds and infrastructures and is very often owned by private industry, but it is becoming apparent over the years that not all developing countries are capable of benefiting from direct aid of this nature. Indeed, one timely change in global perceptions of how North-South relations should develop may be that—except in the case of the very poorest countries that desperately need aid for food, medicine, and so forth—the provision by the North of the sort of aid demanded by many third world countries can be counterproductive in that it perpetuates dependency and often amounts to little more than the transfer of temporary power from one government to another without the creation of anything of lasting value for the recipient country. Much greater lasting benefit would probably result from increased trade between North and South, rather than increased aid from North to South, and from carefully researched efforts to assist developing countries to help themselves by adapting new technologies to the availability of local human and material resources. There surely exists considerable scope for modification of the views previously held by those countries in the "North" still keen to inflict on others the "benefits" of their technology-based societies. Those countries in the "South" that imagine they both need and deserve free handouts of technology and money from the North, but which have not yet sorted out how to cope with them or, indeed, whether they really are both needed and compatible with local cultures and resources, could also well modify their views.

Another change in the international environment relates to the perception that nations have of their own power and the power of other nations. This perception has been made possible largely by the communications explosion of recent decades, but it has been importantly influenced both by the bipolarity of the global power structure referred to earlier and by the dissemination of political awareness stemming partly from the increased role the Soviet Union has been able to play in international relations. It is certainly not the case that nations always have the correct perception or that they are encouraged to develop it, since geopolitically motivated actors are prone to use "disinformation" as well as correct information when attempting to influence the attitude of regimes in their own favor or to the disadvantage of their opponents. However, setting aside the vicissitudes of East-West ideological warfare, Western developed nations in particular are apt to publicize defects in their own systems and to draw attention to those threats from elsewhere that they fear the most.

Of particular relevance in this context is the recognition that, in addition to energy, many of the advanced industrialized countries are already dependent for supplies of a wide range of raw materials upon, in many cases, the developing countries, most of which are considerably less well-off than, for example, some of the oil-producing countries of OPEC. A detailed evaluation of this range of dependencies is beyond the scope of this book, and while the true significance of some of them may be disputed, what cannot be disputed is that, for the time being at least, those dependencies will grow, the pace of the growth depending upon economic growth rates and the ability of technologically advanced nations to come up with alternatives. The question has been raised many times of whether or not the producer countries rich in raw materials could or should combine in some way to exert commercial or political pressure upon the importers, as happened in the case of the OPEC countries. No clear-cut answer has yet been given, although the indications are that they would have a great deal more difficulty in doing so than did OPEC, given the disparity of interests exhibited by those countries and the fact that the formation and cohesion of cartels depends largely upon tight resource markets. Not all resource-rich countries are part of the so-called third world—as the examples of Canada, South Africa, and the Soviet Union clearly show—and there is usually at least one developed industrialized country that has a sufficiently large share of the global production of most of the key raw materials for it to be able, in appropriate circumstances, to minimize the impact of any potential cartel-type activity directed against any particular group of nations. But it does provide food for thought about what is really meant by the "have" and "have not" nations and whether those roles might possibly reverse before long.

The issue of dependency and interdependency is relevant to the unequal distribution of global resources of whatever kind, and any form of dependency is also a form of interdependence involving costs to both sides.[1] The problem is to minimize the costs involved when

1. Robert O. Keohane and Joseph S. Nye, *Power and Independence* (Boston: Little, Brown, 1977).

dependency becomes increasingly asymmetrical, in an era when, for example, the use of military power to resolve differences over commercial matters between strong and weak nations is no longer considered acceptable by the international community. And is there such a thing as true independence? It is worth reminding oneself in this connection that the Soviet Union is very largely self-sufficient in just about all raw material resources and can afford either to stand aloof from the controversies raging or likely to rage around these issues or to involve itself actively in an area that could prove extremely fertile for the nourishment of geopolitical ambitions. The great fear of many people is that increasing amounts of time, money, human effort, and probably human blood will be spent on attempts to determine the ultimate destination of global resources while less and less will be spent upon their efficient mobilization. If "resource wars" come to be fought, they will provide yet another example of how changes in human aspirations outstrip man's ability to adapt his systems to cope with the changes. Those nations that clamor for a new international economic order may come to regret the fact that they did not make greater efforts to see that the existing one worked properly. Similarly, those insisting upon the redistribution of global wealth might think carefully about how that wealth is to be generated and whether they would be happier with a larger slice of a relatively smaller cake or a relatively smaller slice of a larger cake. The countries of the industrialized Western world clearly can give the lead in helping to create wealth for the future, as they have done in the past, but they may have to change their views upon how it is to be distributed and upon the size of their relative shares.

It is unfortunate that the East European countries of the industrialized world seem to be in no position economically to contribute positively toward the resolution of some of the economic problems accompanying the changing international environment. Their political leaders maintain that the development problems affecting the world outside the Communist bloc are entirely of the making of the former colonial and imperialist powers and that it is up to the latter to solve them along lines so often recommended by the great architects of Marxism-Leninism. It is perhaps the best possible indictment of this essentially negative approach that nearly all those developing countries, largely in Africa, that initially espoused the economic courses advocated by the Soviet Union and accepted large-scale East-bloc economic aid subsequently turned away from them whenever they were in a political position to do so. In the longer term, economic conditions in those countries invariably deteriorated and the quality and reliability of East-bloc aid, apart from that in the military and ideological fields, was less satisfactory than that which could be obtained from the West. Many developing countries' perceptions of the suitability of the Soviet Union as an economic partner deteriorated further at the end of the 1970s with the Soviet invasion of Afghanistan, and with East-bloc aid in any case becoming harder to come by in view of the latter's domestic economic problems, another change in the delicate balance of international relations could be said to have occurred. While it would be wrong to consider such a change as irreversible or of outstanding importance, it is a change that could subtly affect the attitudes of the more vociferous critics of the "old" global order. It is also a change that should be noted by those in the West—hawks and doves, the Right and the Left alike—who have a tendency to depict East-bloc achievements on the international scene as those of people who are larger than life.

DISCONTINUITIES AND DIPLOMACY

The changes in the Soviet leadership, which started to occur in November 1982, could have great implications, and it is with a brief look at these with which this chapter on the changing international environment will conclude. Soviet officials often complain that, in attempting to deal with the United States, they have great difficulty in knowing where they stand with respect to one administration or another. Even during the lifetime of one administration, U.S. policies toward the Soviet Union can change dramatically, they maintain, instancing the apparent 180-degree turn in the Carter administration's approach to East-West matters following the events in Afghanistan. They also point to U.S. trade policy toward the Soviet bloc and the imposition or lifting of embargoes on this or that commodity, apparently at the whim of successive presidents. This alleged inconsistency on the part of the United States toward the Soviet Union may or may not be a good thing in terms of overall East-West relations. There are some who would maintain that it is not a bad idea to leave an adversary in doubt about how you are going to react to what he does since it makes it harder for him to outwit you. Others would say that, particularly in times of heightening tensions, this does nothing to build confidence between opponents and may indeed encourage one's adversary to take more risks because he can no longer depend upon having a rational relationship and is likely to lose out whether he behaves "well" or "badly."

What strikes one about changes in the Soviet leadership is that, whatever the personality of the new overall leader, the complexion of Soviet policy does not change abruptly, and it is perhaps relevant that those in the top echelons of Soviet policymaking, and particularly those concerned with foreign policy, tend to be elderly and very set in their ways. It is also difficult to dispute the fact that the broad lines laid down for Soviet foreign policy, and certainly the long-term objectives behind them, have not changed much in recent years even though Western perceptions of that policy have indeed changed or, perhaps, oscillated. Admittedly, progress or lack of progress in trade relations between the Soviet Union and other countries is profoundly affected by the state of political relations between them, since Soviet foreign trade is an integral part of the overall machinery used by the Communist party to further its global objectives. But personalities seem to play a less important role in the formulation of Soviet foreign policy than they do in Western countries, and it might therefore be expected that, initially at least, there will be little change in Soviet foreign policy now that First Secretary Brezhnev has been replaced by Andropov.

It is possible, however, that a discontinuity may be approaching as the 1980s progress, since the age of the "old guard" of those who have occupied leading positions in the Soviet hierarchy for so long is such as to preclude their holding office much longer, and it is questionable whether a new generation of Soviet leaders will be motivated in quite the same way. It is now almost seventy years since the revolution that brought the Bolsheviks to power in Russia and almost forty years since World War II, in which many of the Soviet leadership, whose average age is well over sixty-five, had direct experience. While there is no guarantee that the average age of a new Soviet leadership would be substantially lower, it is probable that many of those succeeding to power will not have directly experienced either their country fighting for survival, as in World War II, or times when it was economically and militarily weak, as in the 1950s and 1960s. Indeed, they will probably be accustomed to a Soviet Union steadily increasing in economic and military might and in international stature, and they will have witnessed the decline in power of the United States and Western Europe. Whether this will mean that they will be less imbued with ideological fervor or more assertive and determined to use their perceived power remains to be seen. It could be that they will be more enlightened and flexible than their predecessors and prepared to try for both a new and more relaxed relationship with the outside world and a restructuring of the domestic economic machinery, which is clearly ailing. They might have to accept that this would necessitate a different political approach in many fields in which their predecessors were unwilling or unable to countenance change. On the other hand, it could mean that, faced with a fragile and potentially explosive situation in Eastern Europe and a deteriorating economic situation at home, they could resort to harsh measures, becoming more repressive than ever in dealing with the East European members of COMECON (Council for Mutual Economic Assistance) and more belligerent in their global activities. It is clear that here is one aspect of the changing international environment which merits a great deal of attention.

The international order at the beginning of the 1980s appeared to be characterized by something of a stand-off between the world's two greatest powers and by a proliferation of instability in most other parts of the world. The so-called nuclear stalemate, which has the effect of neutralizing the ability of either superpower effectively to intervene directly and militarily in any situation outside its immediate sphere of influence for fear of escalation, appeared to have opened the door to numerous lesser powers to indulge in military adventurism. In the developed world as a whole, order, as represented by the absence of war or changes in political alignment, was comparatively stable but with economic problems predominating and the prospect that, one day, access to raw materials might become so important for national survival that conflict might be necessary. In the third world, however, with little to prevent them, local "bullies" could now attempt to settle "old scores," and many new areas for dispute could emerge for political, economic, territorial, or ethnic reasons. Some of these might be virtually irrelevant to the practical interests of the developed countries, but others would be of considerable significance and the temptation on the part of outsiders to "do something" would be great. Perhaps the most fearsome example would remain the Middle East. If one considers that the Soviet Union might also have a practical interest in securing access to long-term supplies of hydrocarbons from that area, then one would indeed have all the ingredients necessary to bring about an uncontrolled, catastrophic change in the international environment.

The changes briefly referred to in this chapter were selected from many others to emphasize that the international climate, in which the negotiations that will determine the extent to which natural gas will become internationally traded over the next few decades will take place, is going to be very different from that in which many of the existing international trade flows were determined in the past. New political approaches can be expected on the part of many producing countries, and new commercial approaches are going to be required by many in consuming countries to cope with them. Unfortunately, there are likely to be further cases where one side, manifesting that it has not adequately adapted to the changing circumstances, will prevent or destroy promising opportunities for furthering the international gas trade, by failing to appreciate the degree of compromise now necessary.

It would seem that any attempt to write an account of the geopolitics of natural gas would be incomplete without a more detailed analysis of the energy situation in the Soviet Union. How likely is it that its geopolitical interest in the Middle East might incorporate strictly commercial elements? It is to this aspect that our journey of investigation will turn in the next chapter.

8 THE EXPORT OF NATURAL GAS FROM THE SOVIET UNION

DISTANCE AND DOMESTIC DIFFICULTIES

Probably no one knows for certain how much natural gas is in place beneath the surface of Soviet territory, but it is believed to be a very large amount. Exploration has not been fully carried out on anywhere near all of the potential hydrocarbon-bearing areas, and even where it has, drilling has only reached comparatively shallow depths, mostly onshore, and the deeper that Soviet technology allows the gas industry to drill the more chance it has of discovering gas rather than oil. Proven reserves of natural gas in the Soviet Union are constantly being revised upward. In the spring of 1982 they were believed to be in the region of 35 trillion m^3, representing over one-third of global proven reserves. Soviet ultimate potential has been put at upwards of 120 trillion m^3. With Soviet natural gas production in 1980 of 435 mrd m^3, and annual increases in the region of 7 percent, a 1982 production of just under 500 mrd m^3 would give a proven reserves-to-production ratio of 70 to 1. There is therefore no expectation that production growth will become resource constrained for the foreseeable future. What may be said already to be a major limiting factor for Soviet natural-gas production is the remoteness of some of the main proven reserves from domestic consumption areas and from export markets. Not only are the reserves primarily located in hostile geographical environments—subject either to the permafrost conditions of, for example, western Siberia or to drought and desert heat in Central Asia—but they are mainly in underpopulated and underdeveloped areas where, often, entire infrastructures have to be created before the enormous task of transportation over thousands of miles can even be embarked upon. It is a remarkable achievement that the Soviet gas industry has managed a tenfold increase in production over the past quarter of a century, and there is little doubt that natural gas is now regarded by Soviet planners as a top priority for investment and development although, in the 1950s and 1960s, the industry suffered from competition for available manpower and material resources with the longer established oil industry. Indeed, with the oil industry facing a leveling off of production levels, and with both the coal and nuclear industries suffering from technological and material shortcomings, natural gas has to play a key role in the provision of incremental Soviet energy supplies during the 1980s and possibly beyond.[1] Not only must domestic requirements be satisfied by greatly extending existing distribution networks and building new ones, but natural gas must also be supplied to other European members of COMECON, which already face considerable energy constraints to economic growth consequent upon the decision by the Soviet Union to limit oil deliveries to these countries at the 1980 levels, at least through 1985. Furthermore, with hydrocarbon exports accounting for over 60 percent of Soviet hard currency earnings in 1981 and with oil export surpluses dwindling, exports of natural gas will play an increasingly important role in Soviet foreign trade generally, assuming that the Soviet Union will have an enduring need to import a wide range of goods and services from the countries of the hard currency area.

Faced with the huge distances involved in bringing Soviet natural gas to both domestic and export markets in the required quantities and within the scheduled time frame, Soviet planners realized in the late 1960s that domestic production facilities for large-diameter steel pipe, compressor stations, and other hardware would not be adequate and that substantial imports would be necessary. The problem was how to finance those imports, given the chronic shortage of hard currency prevailing throughout COMECON. The solution decided

1. Jonathan P. Stern, *Soviet Natural Gas Development to 1990* (Lexington, Mass.: D.C. Heath, 1980).

upon was the export of large quantities of natural gas to Western Europe. Initially, the gas was pumped from fields in the west of the Soviet Union with deliveries confined to the countries of Eastern Europe via the "Brotherhood" pipeline, but in 1968 the first exports to the West occurred under a contract with Austria, and West Germany signed its first Soviet contract in 1970. Subsequently, contracts were negotiated with Italy for deliveries commencing in 1974 and with France for deliveries in 1980. During the late 1970s, the Soviet Union entered into a four-cornered arrangement under which Italy purchased from the Netherlands quantities of gas which were then delivered to France, while Italy received an equivalent amount of Soviet gas which was delivered on French account. By 1980 the Soviet Union was exporting about 23 mrd m^3 of natural gas to Western Europe, of which West Germany was taking 10.7 mrd m^3, Italy 6, France 4, Austria 2.6 and Finland 1.1 mrd m^3 per annum. It was historically an essential part of all Soviet contracts with hard-currency gas-importing countries that deliveries of gas should initially be compensated for by supplies of large-diameter steel pipe or other energy-related hardware, and this is still the basic position today. In the case of West Germany, for example, a gas-purchasing consortium headed by the gas-distribution company, Ruhrgas, receives the Soviet deliveries, while very large quantities, up to 1 million tons at a time, of the pipeline are delivered to the Soviet Union by Mannesmann, a steel manufacturer. Since the contracted gas deliveries take place over a period of several decades whereas the pipeline and other equipment is required in the short term, the latter is delivered on credit terms, involving usually between 1 and 1.5 billion Deutsch marks (DM), coordinated by a consortium of West German banks. Similar arrangements have occurred in respect of Soviet gas deliveries to Austria, Italy, and France.

It is of interest that, for primarily logistical reasons, the Soviet Union does import some natural gas itself into the southern republics, from Afghanistan and Iran. In the latter case the initial deliveries of gas were in payment for Soviet material for and assistance in the construction of a steel plant, and the 1,000-kilometer pipeline, known as IGAT I, was at one time delivering up to 12.5 mrd m^3 per annum of associated gas from the oilfields in the south of Iran. When the war between Iran and Iraq flared up in the late 1980s, and following the domestic turmoil resulting from the overthrow of the Shah, Iran's oil production fell dramatically from the 6 million barrels per day achieved during the later 1970s, and deliveries of natural gas to the Soviet Union were thus sharply curtailed. This caused significant logistical problems for Soviet planners, who were obliged to divert alternative, mainly liquid, fuels from northern and western parts of the country to keep the southern republics functioning during the particularly heavy winters of 1979/80 and 1980/81. It had been planned that, through a second deal with Iran—IGAT II—additional volumes of Iranian gas would have been delivered to the Soviet Union which, after deducting a quantity in lieu of transit fees, would have delivered a similar volume: 11 mrd m^3 per annum to Austria, West Germany, and France. While some quantities of Iranian gas were believed to have been delivered to the Soviet Union in 1981/82 via IGAT I, the idea of IGAT II was shelved indefinitely after the revolution in Iran.

Domestic consumption of natural gas in the Soviet Union, which was running at about 430 mrd m^3 per annum in 1982, has never been resource constrained, but, largely as a result of the enormous distances involved in bringing gas to the markets, by no means all potential consumers have yet received supplies, and demand is therefore unlikely to be satisfied for many years to come. The problem is primarily one of constructing both major long-distance carrier lines and a network of shorter distribution networks on a very large scale. There are also believed to be bottlenecks in the production and supply of various types of outlet burners for both commercial and domestic uses. Consumption increases may, for these and other reasons, not be able fully to keep pace with production increases, and the surplus potentially available for export may therefore be expected to increase, at least during the 1980s. Clearly, the development of the domestic network is linked with the fortunes of the export market and, thus, to the ability of the Soviet Union to import large-diameter steel pipe and related equipment for the gas distribution industry, such as compressor stations and valves. To the extent that the Soviet government is encouraging, on the one hand, the conservation of oil use and, on the other hand, increased use of natural gas, it is important that the development of the domestic gas market proceeds smoothly and rapidly, thereby freeing up as much oil as possible for export.

Domestic capacity to produce steel pipe of the required diameter and quality has constantly lagged behind demand, and there are a number of other areas in which imports from non-COMECON sources are important for the smooth and optimum rate of development of the Soviet energy industries in general and the natural gas industry in particular. However, it would probably be a mistake to assume that without such imports these industries would be in serious difficulties even though shortcomings in their overall performances undoubtedly adversely affect the chances for fulfillment of the overall Soviet Plans. In this connection it may be mentioned that Sovietologists in the West tend to set a great deal of store by the extent to which the Soviet Union fulfills or underfulfills its annual and five-year or longer term plans, but, particularly in the case of the energy industries, production targets are usually set within quite a wide range, and there may be several motives behind the selection of the ranges that ultimately become official. While the plans undoubtedly do carry great weight in most Soviet circles and are of considerable exhortatory value, there is a danger that too much significance may be given by outsiders to the extent by which they may be achieved or underachieved and thus to the inferred degree of dependency upon outside assistance that may be indicated.

The ability of the Soviet industrial system to accommodate shortcomings in various of its individual sectors is considerable, mainly due to the proven ability of the Soviet consumer to go without more, and for longer, than would probably be tolerated by Western counterparts. In the case of the development of the natural gas industry, for example, there is little doubt that Western assistance on a substantial scale would be of considerable benefit if it could be achieved on favorable commercial terms and without the need for the payment of political premiums. But this is far from saying that the development could not proceed in the absence of such assistance or that the Soviet consumer would be in a position effectively to protest in the event that deliveries of gas were delayed as a result of any political decision to forego such assistance if the price were to be considered by the Soviet government to be unacceptably high.

In summary, it is clear, first, that there is little likelihood that the very large gas export capability of the Soviet Union will be constrained, over the next decade at least, by either resource availability or a rapid growth in domestic demand. Second, there are strong and enduring incentives for the Soviet Union to maintain and increase its exports of natural gas to the hard currency areas, and particularly to Western Europe. Third, it would be wrong to assume that any imposition of a ban on Western technological or material assistance would significantly harm, let alone cripple the development of the Soviet gas industry. And fourth, the auspices would currently seem to be most favorable for potential Western purchasers of Soviet natural gas to enter into long-term contracts with that country on purely commercial grounds.

COMMERCIAL ASPECTS

It is appropriate here to consider in more detail some of the commercial aspects of the East-to-West flow of natural gas on the European continent, although, in passing, it should be noted that possibilities do exist and are indeed under negotiation for Soviet gas exports from the far eastern part of the country, with the potential purchasers in this case being the Japanese. Perhaps the first thing that strikes one is the obvious availability of Soviet gas, in spite of the immense distances across which it must be piped before it reaches the borders of Austria and Germany. A major pipeline system has already been in existence for a decade. This is the "Transit line" across Czechoslovakia, and it connects with two main east-to-west systems within the Soviet Union, already able to supply over 50 mrd m^3 for export and capable of increased throughput with comparatively little additional capital investment. In other words, the mechanism is already in place to deliver substantial additional volumes of natural gas to Western Europe within a very short time scale, and it is capable of considerable expansion to accommodate increases in both demand and supply. Of course it is not just a question of Western European countries being able to purchase the gas at the border for cash. The Soviet Union is concerned with ensuring that its exports of this depletable hydrocarbon resource result in the acquisition of hardware and equipment that will add to the infrastructural development of the Soviet hinterland. If Western importers want the gas, they must be prepared to contribute rather more than just money toward its availability, particularly since the ideological and organizational differences between the capitalist and communist systems currently preclude the forms of investment in raw-material acquisition from Soviet sources that would be considered normal in many other parts of the world. Western companies are not, for example, permitted to go in to explore and produce Soviet natural gas on the same risk basis that they are in other places, and the Soviet State has complete control over the production and disposal of its natural gas, preferring to deal on a government-to-government basis wherever possible. Nevertheless, and particularly with respect to offshore exploration and production, there exists a surprising degree of flexibility on the part of, at least, those Soviet officials directly involved in the detailed negotiations entered into with Western companies, and if, miraculously, political and ideological considerations could be excluded, there is every likelihood that Soviet natural gas resources would be quite rapidly developed for the benefit of both Soviet and foreign consumers.

Another fundamentally important commercial aspect of the future for Soviet natural gas exports to the West is the price at which buyers in Europe can obtain it, and we have seen how, during the 1970s, European import prices were largely determined by the availability of low-priced Dutch gas. Even in the spring of 1981, Soviet gas exports to, for example, Western Germany achieved only 77 percent of parity with imported crude oil prices, while West Germany was paying less at that time for its Dutch and some of its Norwegian gas. To be more specific, in April of that year, crude oil was imported at DM 615 per ton, and with the exchange rate then DM 2.30 to the U.S. dollar, Dutch gas was imported at $3.7/MMBtu, Norwegian at between $3.20 and $5.10, and the existing contracts with the Soviet organization Sojuzgaz-export were averaging $4.6/MMBtu. Algeria's campaign to secure export prices similar in terms of energy equivalence to those it was seeking for its crude oil exports may have influenced the Soviet negotiators, since, for the negotiations associated with the new Yamburg-Urengoi Project, they initially asked for 100 percent parity with low-sulphur fuel oil, the equivalent of 95 percent of crude oil parity at the time, which would have been $5.7/MMBtu. As it turned out, the new contract was eventually concluded toward the end of 1981 at a price reportedly in the region of $4.70 to $4.80/MMBtu, although it is never particularly clear (from deals in which one of the parties involved is a sovereign power that expressly requests that full commercial details not be made public) exactly what a given price actually means. One thing that was clear however, was that a new deal between West Germany and the Soviet Union would simply not have taken place, for purely commercial reasons, had the latter insisted upon the higher price. It should be stressed at this point that the negotiations between the West German buyers, led by Ruhrgas, and Sojuzgaz-export were totally commercial, and no attempt was made by West German politicians to influence the final outcome of what was reportedly a very tough commercial struggle. Similarly, in spite of the shareholdings in the West German buying organizations held by some of the major international oil companies, the latter had no majority position and would have been able to exert only limited influence one way or the other even if they had felt that too much Soviet gas flowing to Western Europe might adversely affect their interest in the development of gas projects elsewhere. West Germany has always been the key country in respect of West European imports of Soviet gas, and it is quite likely that had it not been persuaded to take a substantial quantity, 10.5 mrd m^3 of the envisaged 40 mrd m^3 annually, the project might well have foundered, so the Soviet Union had to take a lot of things into account when deciding to reduce its asking price, and by no means all of these were strictly commercial.

It has to be understood that the project to export gas from, initially, the Urengoi fields and subsequently from those on the Yamal peninsular to Western Europe, which will henceforth be referred to as the Yamburg-Urengoi Project, was primarily a domestic project aimed at developing the substantial northwest Siberian deposits by means of a network of six trunklines. Only one of these lines could claim to be an export line in that it alone was to be constructed over a distance of 5,000 kilometers in a southwesterly direction to join up with the existing export-oriented lines converging on the Soviet-Czech border near Uzhgorod. The cost of this 56-inch line was put at about $10 billion, of which about $4 billion was intended to pay for equipment supplied by the West and financed in whole or in part by the suppliers and their banks. The $4 billion was to be credited directly to the pipeline and equipment suppliers and would at no point be at the disposal of the Soviet Union, which would, however, have to bear the local costs amounting to approximately $6 billion.

It is expected that the provision of the key hardware to be supplied by, for example, West Germany, amounting to some $1 to 2 billion in value, will ensure employment for not more than a few thousand workers over the three-year construction period—an important number, but not large in relation to a total work force of 23 million, of which some 100,000 have been estimated to depend upon continuing overall trade between West Germany and the Soviet Union. Although there is no doubt that Western European governments welcome the employment and plant utilization that the Yamburg-Urengoi project represents, its significance in this connec-

Figure 8-1. Routes of Soviet Gas Exports for Western Europe.

The Energy Journal 3, no. 3:19.

tion should not be overestimated. The building of the other trunklines for the development of the Urengoi field is almost entirely a Soviet and COMECON-financed affair although imports of large-diameter pipeline and compressor stations are certainly going to be required for their construction too. The project can be seen as a sophisticated mechanism to enable the development of Urengoi, for domestic purposes, to proceed with the assistance of those West European countries that are facing potential natural gas supply shortfalls toward the end of the century. The project appears to make a lot of commercial sense to both the buyers and the sellers, and if it does not come to fruition, for whatever reason, the development of the Urengoi field, already producing over 80 mrd m^3 in 1982 and scheduled to produce over 230 mrd m^3 by the mid-1980s, will undoubtedly be delayed somewhat, but probably not by more than a year or two. This would assuredly have some adverse repercussions for the planned Soviet energy output targets, but the West European buyers would then face considerable problems in ensuring, by the second half of the 1980s, replacement contracts for adequate natural gas imports from elsewhere even though, in the longer term, sufficient quantities would very likely be forthcoming from Norway, Nigeria, and other sources.

It has been argued that the constant and long-term availability of large volumes of Soviet pipeline gas for Western Europe is a disencouragement to those governments and other organizations considering involvement in supplies from other sources, for example, by pipeline from the Norwegian North Sea reserves or by liquefied natural gas (LNG) delivery from West Africa or Canada. It certainly is the case that supplies from the Soviet Union are highly convenient and that, provided the price is competitive, there can be few commercial reasons why a significant proportion of West Europe's imported gas should not come from that source. Indeed, if solely the financial interests of the West European consumers are considered, the availability of Soviet natural gas must be a positive contribution to the reduction of overall energy costs. From a purely commercial standpoint it could further be argued that it is up to other potential suppliers interested in the West European market to ensure, if they can, that their prices are also competitive with whatever it is that the Soviet competition has to meet, such being the nature of a buyers' market.

Somewhere along the line, any producer country will have to decide whether or not it makes better commercial sense to compete in a particular export market for natural gas or to leave the gas in the ground or to use it domestically. The Western European market is not a homogeneous affair, and there are several different markets, both from a geographical and sectorial point of view, some of which are not in the running for Soviet gas in any case. Furthermore, it is not good commercial practice for any market to let itself become overly dependent upon any one supplier. Even if large volumes of Soviet natural gas are likely to continue to move to certain West European countries, those countries will seek other suppliers as well while other West European countries may continue to prefer other suppliers to the Soviet Union. Of course, the netbacks to producers will not always be similar and some will tend to do much better than others, but it seems that, while the attractiveness of certain potential supply schemes for Western Europe may be reduced in some respects as a result of the Yamburg-Urengoi Project, it can be enhanced in other respects, and it must surely be in everyone's interest that sources of supply for Europe should be mobilized in as many places and as often as possible. Experience in the early 1980s with the pricing of natural gas exports from Algeria has indicated that relying on too few sources in the international natural gas business is not advisable, and Western European buyers must consider themselves most fortunate to have a number of substantial potential suppliers with which to negotiate as their main local continental source approaches the end of its export capability. It is up to the West Europeans to ensure that those alternative suppliers find it worthwhile, given their various political and commercial idiosyncrasies, to remain in contention, and this also applies in the case of the Soviet Union.

One final commercial consideration must be related to the need to conserve domestic energy resources in Western Europe wherever possible. If additional volumes of Soviet natural gas were not going to be available to the West European buyers in the second half of the 1980s, the pressure on domestic resources would have to increase, given the long lead times required to bring new export projects on stream. But once those domestic resources have been depleted, reliance upon imports increases even more, if natural gas is to maintain its share of the fuels/energy balance. Again, it seems logical that West Europeans with static or declining domestic production should buy natural gas from a country like the Soviet Union while it is still anxious to sell. In this way the drain on domestic reserves can be reduced, prolonging somewhat their availability and conserving them for use in a time of, for example, disturbances in the global energy supply situation. It is not inconceivable that, whether for political or other reasons, the Soviet Union might one day be less willing or able to sell natural gas to Western Europe than it has shown itself to be in the 1970s and in the early 1980s, and other suppliers would be able to take advantage of such a situation, particularly if domestic reserves on the continent were, by that time, that much lower. Of course the availability of such plentiful supplies as those offered by the Soviet Union should not preempt the search for and development of domestic resources in Western Europe. This can sometimes happen if nations do not develop clear-cut policies concerning the provision of subsidies or the creation and maintenance of reserves.

Before considering various noncommercial aspects of Soviet natural gas exports it is interesting to note that many of the large orders for machinery and equipment that West European suppliers had been led to believe would materialize once the new purchase contract for Soviet gas was signed had not, by the summer of 1982, materialized. There was, as a result, considerable disillusionment in business and financial circles, particularly in the Netherlands and West Germany. It was felt that those who had pressed so hard for signature of the contracts had been misled by Soviet negotiators. Largely as a result of this, it did not seem likely that the Netherlands would take anything like the 3 to 4 mrd m^3 of Soviet gas originally discussed under the Yamburg-Urengoi Project, and might not take any at all, particularly if Soviet orders for machinery and equipment were not going to materialize. The status of the Yamburg-Urengoi Project in April 1982 was that West Germany had signed for 10.5 mrd m^3, with a further 0.7 mrd m^3 agreed upon for delivery to West Berlin, and France had signed for 8 mrd m^3. Italy was on the verge of signing for 8 mrd m^3 but was still evaluating its position for both political and commercial reasons. In total, therefore, 19.2 mrd m^3 was at that time scheduled to flow under the project eventually, following start-up in 1984/85. Austria and Switzerland signed during the summer of 1982, and if Italy, Belgium, and Spain should also sign, the maximum could rise to nearer 35 mrd m^3 by the end of the 1980s, although the originally envisaged volume of 40 mrd m^3 seems unlikely to flow until the early 1990s. The project has clearly been modified in both throughput and timing and may yet be further modified should political or economic circumstances require it. In this connection it is believed that, for example, the contracts that have so far been signed contain clauses that would allow buyers to obtain up to a 20 percent reduction in purchased quantities over a given period should a fall-off in domestic demand necessitate it. It should be borne in mind that total Soviet deliveries to Western Europe under existing contracts amounted to some 23 mrd m^3 in 1980, when overall consumption was 208 mrd m^3, but that these are due to decline to nearer 16.5 mrd m^3 by 1995 unless extensions to the contracts are negotiated. If the maximum project volume of 40 mrd m^3 is flowing by that time, overall West European imports from the Soviet Union would be unlikely to exceed 56 mrd m^3 in the absence of new deals, and consumption in that year might be in the region of 270 mrd m^3. Initial deliveries under the project will, in fact, probably not be of Urengoi gas but of gas from the Orenburg field supplied via the Soyouz pipeline, which currently has spare capacity. Urengoi gas will be phased in during stage two, while the Yamburg gas will not flow until, under the third stage, a 185-mile connecting pipeline is built to join the Urengoi line.

However and from whatever source the gas reaches Western Europe, the significance of the Yamburg-Urengoi Project is that Soviet natural gas will increase its share of West European natural gas consumption from some 11 percent in 1980 to nearer 22 percent in the mid-1990s. In 1980, with natural gas accounting for 14.5 percent of total West European energy consumed, Soviet gas represented 1.6 percent of that total. In 1995, even if 60 mrd m^3 of Soviet gas were being imported and if natural gas should achieve a share of 16 percent of total energy consumption, imports from the Soviet Union of this hydrocarbon seem unlikely to exceed 3.5 percent of that total. In view of the fact that many West European countries, notably Portugal, the United Kingdom, Ireland, Norway, Sweden, and Turkey, do not now take, and are unlikely in the foreseeable future to take, Soviet natural gas, it is clear that, in some of those countries that do, the percentage of overall energy and natural gas consumption accounted for by Soviet deliveries will be substantially higher. Take the case of West Germany, which imported 10.7 mrd m^3 of Soviet gas in 1980 and consumed 52 mrd m^3 in total. In that year Soviet deliveries accounted for nearly 18 percent of gas but less than 2.8 percent of overall energy consumption. In 1995, while it is most difficult and perhaps imprudent to make detailed predictions, it seems that the likely maximum of 20 mrd m^3 imports from the Soviet Union could represent up to 35 percent of total gas consumption and between 5.5 and 6.5 percent of total energy consumed. For France, Soviet gas accounted for 15 and 1.7 percent, respectively, of total gas and total primary energy consumption in 1980. For Italy the comparable figures were 24 and 3.6 percent. By 1995 Soviet natural gas might account for over 32 and approximately 20 percent, respectively, of French and Italian gas consumption. For Austria, the 1980 and 1995 estimates of the gas consumption share accounted for by Soviet natural gas would be 60 and approximately 80 percent. The question arises whether these percentage figures are significant within the context of the security of the countries' existing and future energy supply, and this leads into the whole complex of political aspects of East-West relations.

TRANSATLANTIC TENSIONS

The Soviet Union had been exporting natural gas to Western Europe since the early 1970s and had even been a party in negotiations for a triangular deal involving, in effect, the export of Iranian gas to Europe via the Soviet pipeline system. Yet, by the end of the decade, there had been no particular outcry or wave of hostile sentiment against the principle of Western Europe taking Soviet gas. However, the invasion of Afghanistan was undoubtedly a catalyst for change in the perception of Soviet behavior by many countries around the world. Had the fact that negotiations between the Soviet Union and West European natural gas buyers over a major new gas export project were underway been more widely known at the time of the Carter administration's reaction to Afghanistan in 1980, it is likely that the project would have come under open attack even earlier than it did. But it was not until the Reagan administration had taken office and had already manifested a considerably harder line toward what it considered to be a new phase in Soviet expansionism that the issue became prominent in East-West relations, particularly following the events in Poland.

By the autumn of 1981, the U.S. administration was set upon preventing the project from proceeding and tried extremely hard to persuade the West Europeans to drop the idea on the grounds that its realization would make "Western Europe" overly dependent upon energy imports from the Communist bloc and thus subject to political leverage from Moscow over strategic and security matters. It was also felt in Washington that the granting of large volumes of long-term credits to the Soviet Union at favorable rates—and the provision of large amounts of large-diameter pipeline, compressor stations, and related hardware under these credits—would significantly strengthen the struggling Soviet economy and enable the Soviet leadership to continue building up the country's military strength and avoid having to divert investment effort from the military to the ailing civilian sectors.

This extremely high-profile approach on the part of the Reagan administration produced what, for the hard liners in that administration, must have been an extremely disappointing response from the West European countries actually involved in negotiations over the project, and although the project as originally envisaged came for various reasons to be modified over the period of most intense U.S. opposition, two out of the three key importing countries—West Germany and France— had signed contracts with the Soviet Union by early 1982. By the spring of 1982 it had become clear to the administration that its efforts to kill the project were not likely to succeed unless there were a further deterioration in East-West relations. President Reagan then made it clear in June of that year, through certain licensing restrictions that the United States imposed upon the export of components and technology for some important items for the project, that whatever else the West Europeans might do, the United States was not prepared to conduct business as usual with the Soviet Union, especially following the imposition by the authorities of martial law in Poland, for which the administration held the Soviet Union largely responsible. This embargo, which would not be lifted until the ending of martial law, seemed capable of somewhat delaying the actual completion of the project although there seemed little doubt that the Soviet Union would, as it lost no opportunity of proclaiming, be able to commence deliveries to Western Europe by the end of 1984, albeit largely thanks to spare capacity in the existing export network.

For about eighteen months the issues raised by the Yamburg-Urengoi Project exacerbated not so much East-West relations as relations between the Western allies. It highlighted many disturbing divergences of interest between those members of the Western Alliance finding themselves on opposite sides of the North Atlantic. What was seen by West Europeans as a simple extension of a mutually beneficial commercial relationship that transcended the ideological barrier across Europe appeared to some in the United States as a betrayal of the Western cause and a fundamental strategic error that would have long-term and far-reaching adverse repercussions. These in turn would irretrievably weaken the alliance and strengthen the hand of the Soviet Union in pursuing its expansionist aims. How such a divergence of views could have come about is not easy to explain, but there follows an attempt to clarify some of the issues involved and to emphasize a number of points that seem to be most important in arriving at an understanding of the true significance of the Yamburg-Urengoi issue.

DEPENDENCE AND DIVERSIFICATION

It is necessary to understand that, for reasons referred to elsewhere, the dependency of most West European countries upon imported energy is perforce considerably greater than that of the United States, as is their need to diversify away from hydrocarbon imports from OPEC in general and from the Middle East in particular. However, there are very long lead times involved in all of the obvious diversification options, and while increasing use may indeed be made of nuclear energy and imported coal, hydrocarbon imports will be needed to supply the bulk of West European energy requirements for the foreseeable future, and there are not all that many non-OPEC sources to which Western Europe can turn in the short term. In the case of natural gas supplies, Western Europe is committed, in view of the substantial infrastructure devoted to the distribution of natural gas, to rely on this energy source to meet around 15 percent of its overall energy requirements at least until the next century. However, all existing or potential sources are subject to some degree of political risk and geographical or technical constraints, and this places a premium on those to which commercially attractive access can be achieved in the shortest time. It would therefore be remarkable if Soviet natural gas, with the import of which the West European importers have already had a decade of experience, were not considered an attractive alternative, given its immediate availability and the fact that the Soviet Union needs to expand exports of it.

Although it has been argued by some that the Soviet Union is not necessarily a reliable energy supplier and is liable to use any dependency upon its exports as a political lever—through the threat of a cut-off of deliveries,—this has so far not been Western Europe's experience of large-scale energy supply relationships with that country. It has, on the other hand, experienced such problems with, for example, Algeria. For West Europeans there seems no prima facie reason for regarding the Soviet Union as any less reliable an energy supplier than anywhere else, given that it is not possible to imagine any totally risk-free supply source of imported energy in these times of troubled international relations. At worst, the Soviet Union could be considered just another energy source over which to spread the supply risk, but there are many Europeans who would consider it rather more favorably, taking into consideration what they see as the very real commercial vested interest that the Soviet Union has in proving itself a reliable trading partner.

The United States, on the other hand, has never had a particularly large or solid trade relationship with the Soviet Union and, while there are perfectly reasonable geographical and logistical reasons for this, it must be admitted that what trade there has been in recent decades has been bedeviled by politically motivated interruptions and embargoes of one form or another. It has been particularly confusing to many Europeans that the United States urges its West European partners to refrain from buying Soviet natural gas while, at the same time, the United States finds it acceptable to make large-scale sales of wheat and other feed grains to the Soviet Union. It is of course known in Western Europe that political pressures from U.S. farming areas make such a course of action politically expedient for the U.S. administration, but it is difficult to convince many West Europeans that this access to such a vital commodity as grain, which the United States now persists in granting to the Soviet Union, is not just as important an element in assisting the Soviet economy as is the hard currency and hardware that the Soviet Union would acquire through its sale of natural gas to West Europe. It is argued by U.S. opponents of the gas deal that sales of U.S. grain to the Soviet Union result in a net outflow of hard currency since the grain is totally consumed and ongoing purchases have to be made, whereas the hardware imports from the gas deal result in a permanent enhancement of the Soviet infrastructure and are in any case secured, on credit, by the export of a naturally occurring, albeit depletable, raw material that might otherwise have a less vital role to play in the development of the Soviet economy. It is not clear whether or not the economics of this argument hold water if a long-term view of the issue is taken, but it does suggest that there were those in the Reagan administration in the early 1980s whose attitude was influenced more by a desire to harm the Soviet economy than by a desire to recognize the benefits that might accrue to the West European energy picture and thus, ultimately, to the strength of the Western Alliance from continued commercial cooperation with the major industrial power on the European continent.

In 1981 and 1982, it appeared to many West Europeans as if the Reagan administration, having rejected what could be regarded as the total failure of the policy of appeasement toward the Soviet Union practised by previous administrations and by most West European governments, had gone over to a particularly hard-line approach, one objective of which was to punish the Soviet Union for what it did in Afghanistan and had done or might be about to do in Poland, and another objective of which was to discourage it from further adventurism on the international stage, particularly in the Middle East. While many in Western Europe would support this last objective, they would not necessarily agree that punishing or attempting to punish the Soviet Union by means of unequally and inconsistently applied embargoes of one sort and another is the best means of achieving it. There seems to be a wide measure of transatlantic disagreement over the question of just how important trade weapons really are when it comes to dealing with a country the size of the Soviet Union which, apart from being a military superpower, is self-sufficient in most if not all important raw materials and natural resources. Perhaps as a result of their much longer and closer trading relationship with both the Russians and the Soviet Union, West Europeans are more aware than their American allies of the great resilience of the Soviet people and of the comparatively small dependency upon foreign trade of the Soviet economy as a whole. West Europeans can perhaps better appreciate the determination of Soviet leaders to avoid economic dependency whenever possible, particularly if that dependency is subject to politically motivated pressures aimed at influencing the leading Communist regime in directions contrary to those that it has set for itself. This issue is relevant to the question of whether it is ultimately beneficial to the "West," however defined, to restrict the sale of technology, know-how, and equipment to the "East."

One thing that can be said in this connection is that no Western policy of restriction has much hope of success unless all Western countries capable of supplying the relevant technology and so on agree to adhere to it, and herein lies one fundamental weakness, in this particular context, of the capitalist system, which positively encourages free-market competition among its members. It is debatable whether or not there are certain civilian sectors of the Soviet economy the development of which is actually being held back as a result of, for example, the embargo on various strategic items of technology imposed by the coordinating committee of NATO known as COCOM. It is probably the case that certain developments could proceed much faster if it were not for the embargo, but it should not be forgotten that, in nonmilitary fields at least, the level of sophistication in many parts of the Soviet economy is different from that generally found in the West and that certain sectors have some difficulty in fully assimilating the most up-to-date Western technology when it is commercially available. To put it another way, it would be just as unlikely that a Soviet energy exploration and production venture could operate effectively in the technological and managerial environment obtaining in, say, Texas, as it would that a Texas-trained unit could operate under the Soviet economic system in, say, western Siberia, and there would appear to be considerable dangers in adopting too subjective an approach when evaluating the expected repercussions of this or that potential punitive policy aimed at the Soviet economy. This is not to say that no such policies could be effective, but they would be extremely unlikely to be so if pursued by only one Western country, however powerful. They would be even more unlikely to be effective if other countries not only did not go along with them but actively sought to benefit from their application, for example by securing additional business denied to a country imposing embargoes. The ability of the Soviet Union to play one Western country off against another and one Western company off against others has always been a strong point in its favor, and the case of the Yamburg-Urengoi Project is clearly one where even the Soviet involvement in Afghanistan and, at least up to the autumn of 1982, Poland was not sufficient to unite member countries of the Western Alliance in a coherent policy of trade restrictions against the Soviet Union. That this was likely to happen could probably have been ascertained well in advance had anyone felt it sufficiently important to try. Indeed, had they done so, many of the tensions to which the divided efforts of the transatlantic allies subsequently gave rise would have been avoided.

ENERGY SECURITY AND GEOPOLITICS

The saga of the Yamburg-Urengoi Project to date has served to highlight not only the transatlantic differences over commercial interests in and perceptions of East-West trade in general, and energy trade in particular, but also the growing divergence over security issues. This must stem partly from the fact that the Reagan administration appeared to view East-West trading relations as part of the fundamental global confrontations between good and evil. By contrast, the West Europeans, no longer actually in a position to play a global strategic role either as a united Europe or as individual countries, preferred to consider, for example, the purchase of Soviet gas as a purely local and commercial matter contributing little to the grand strategic design and possibly even capable of ameliorating rather than exacerbating East-West tension over the longer term. It is possible that the administration felt that the West Europeans were being too casual about global security issues or that they were just not aware of the "lions in the path" that were so obvious from across the Atlantic. It is also possible that the United States felt that, since it alone had the potential military capability of confronting Soviet expansionism wherever it occurred, it should be responsible for neutralizing what it perceived to be incipient, creeping, or "back-door" expansionism in Western Europe, by means of which the Soviet Union might achieve neutralization or even control of West European political will through the fostering of energy dependency in certain key countries. In any case, following the oil shocks of the 1970s the question of energy and security was very much to the fore in the minds of American geopoliticians, and it is probable that the prospect of the Soviet Union apparently maneuvering itself into a position of securing an energy "stranglehold" not only upon gas supplies for "Western Europe" but also upon the oil fields of the Middle East, consequent upon its invasion of Afghanistan, was what finally prompted the Reagan administration to get involved in what was, by most criteria other than geopolitical, very much a European affair.

Yet another dimension in which there seems to be a discrepancy between West European and American views is the conception of the ideological threat posed by communism in general and by the Soviet Union's attempts to extend its own special brand of communism in particular. West Europeans tend to be less impressed by the historical record of Soviet efforts to convert other countries to communism than does the United States. Moreover, they do not see the hand of communism behind every significant event worldwide that appears to have adverse repercussions for Western interests. Not only do West Europeans increasingly see economic and other forms of nationalism as more prevalent and more powerful than communism as a driving force in world affairs, but they are perhaps becoming less sure than they used to be both about what is actually meant by "Western interests" and about the ultimate and inherent "correctness" of Western values. They would probably almost unanimously agree with Americans that the values of communism are inherently "incorrect," but, for that very reason, they probably would say that it would stand increasingly less chance of influencing people if it were not constantly given such prominence as "the only viable alternative" to capitalism. In short, West Europeans are less prone than Americans to paint the Communists larger than life, and they point to the ideological and economic problems that appear to be growing within COMECON itself. They might further suggest that the more the systems of the East and West can be brought into nonmilitary confrontation, for example, through increased trading ties, the more obvious would become the shortcomings of the Communist system. Indeed, if the capitalist system cannot compete with the Communist system in those activities, such as trading, that the capitalists ought to know best, for fear of being beaten at their own game, then perhaps there really is no hope of peaceful coexistence on this planet. But the only alternatives seem to be armed confrontation or an attempt at the isolation of one system by the other—an attempt that, for classical geopolitical reasons, would inevitably fail. The Yamburg-Urengoi Project is, in this connection, seen by West Europeans as one more way of opening up and maintaining a manageable channel of communication and cooperation between two major human groupings inhabiting the same land mass, and the fact that it brings economic benefits to both sides is not seen as being to its disadvantage. So the West Europeans tend to stress what they see as the positive aspects of the project, emphasizing the mutual interest that it represents rather than the ideological dangers and direct security risks it could entail.

On the question of the security or lack of security of the project, it must be stressed that the West Europeans were cognizant of the security elements even before the Americans turned the spotlights on them, but there is no doubt that this attention from the United States has had the positive effect of strengthening the West European effort to take appropriate measures in this direction. The issue has been well and truly aired, and there can be few politicians, administrators, or businessmen who are not now fully aware of what has to be done to reduce the problems that West European consumers might face in the event of an interruption of supplies of Soviet gas for whatever reason—political, climatic, or technical. The West European high-pressure gas distribution network is an extensive one, stretching from the North Sea across to the Mediterranean, and it should be remembered that Soviet supplies to any one West European country could not be interrupted without affecting supplies to other importers. Furthermore, if any single country on the network were to suffer disruptions, particularly if these were politically motivated, it would in most cases be possible to arrange substantial back-up supplies, not only from the major remaining reserves in the Netherlands but also from reserves in other countries and from storage capacity that has been and is still being built in many countries. Delivery capacity is usually the major bottleneck in supply disruptions, and the main question here is: How long could those back-up supplies last? This raises the question of how long it would be likely that, in the event there were an interruption of Soviet supplies, such an interruption would last. Given that a long-term cutoff of supplies could not be targeted upon one importer in isolation, if the Soviet Union were to take such a step it could only be the case that it had decided to offend all the major continental powers at the same time. It is difficult to imagine this occurring unless Moscow was prepared to forego any chance of future East-West technical, commercial, or diplomatic cooperation for a long time to come; in other words, unless it was prepared for a war situation. Any short-term cutoff that it could not be demonstrated was caused by circumstances beyond Soviet control would merely serve to irritate West Europeans generally and embarrass left-wing forces there. Such a short-term supply disruption could in any case be coped with under normal back-up arrangements provided it only lasted a few weeks. Anything longer would be tantamount to the declaration of open hostilities by the Soviet Union, and in such a case West European economies would be having more serious problems to contend with than the loss of up to a maximum of 7 percent of their overall energy supplies.

What would be far more serious for West European energy security would be if an interruption in Soviet natural gas and oil supplies were to coincide for some reason with disruptions in other hydrocarbon supplies—for example, in the event of a widespread OPEC oil embargo. But it is difficult to imagine circumstances under which a cutoff of both hydrocarbons could occur by accident or be sustained by design, since the geopolitical interest of the Soviet Union and the majority of oil-exporting nations are not too likely to coincide. Indeed, any attempt by the Soviet Union to take political advantage of the imposition of, say, an OPEC oil embargo on Western Europe by cutting off its gas supplies would probably have the immediate effect of ending the embargo. Once again, it appears that the security risk of a direct and deliberate cutoff of Soviet natural gas deliveries to the Western Europeans would be very small except in circumstances likely to indicate a global flare-up. What has not yet been discussed is the possibility that, at a much lower level of exposure, the Soviet Union might use the threat of a cutoff of natural gas supplies to attempt to influence some imminent political decision in a West European country over, for example, the stationing of a new generation of strategic or tactical weapons upon that country's soil. Unfortunately, those who have put forward this suggestion have never spelled out just what they mean, and it is not a suggestion that meets with a lot of support in many West European circles.

By the end of April 1982 it seemed as though the Reagan administration had virtually abandoned its attempt to kill the Yamburg-Urengoi Project solely on the grounds that it was bad for Western Europe's energy security or political well-being, even though the embargo on the supply of certain items of technology and equipment by U.S. companies for the project remained in force and was subsequently to be extended to cover the activities of overseas subsidiaries of those companies situated in, for example, Western Europe. The attention of those in the United States opposed to the project was turned, first, to trying to persuade Western governments and financial circles not to grant, at least for a while, credit terms to the Soviet Union more favorable than those that would be normal for an industrialized country, and for this there seemed likely to be a rather wider measure of support, and second, to attempting to delay the project by making it harder for West European companies to deliver some crucial hardware. This was, by contrast, threatening to exacerbate U.S.-West European tensions. It may be recalled that the United States had proposed that more sensible alternatives for the diversification of West European energy supply sources would be the purchase of substantially increased quantities of United States coal and the rapid mobilization of the very large natural gas reserves recently discovered in the Norwegian sector of the North Sea. On the first alternative, while imports of U.S. coal into Europe had been increasing rapidly in recent years, the feasibility of boosting deliveries to a level equivalent to the energy content of the envisaged Soviet natural gas deliveries was considered, after further examination, to be very low, due not only to bottlenecks in the supply system on the Western side of the Atlantic but also to insufficient capacity in Western Europe to process greatly increased volumes of coal. Both problems seemed capable of solution in time, but not quickly enough to eliminate the need for increased deliveries of Soviet natural gas in the mid-1980s. On the question of the Norwegian reserves, it was apparent that even if the Norwegian government could be persuaded to agree to their mobilization at a rate far greater than intended, the technical difficulties inherent in developing these particular deposits, in the notoriously hostile environment of the North Sea, would preclude substantial increases in Norwegian supplies of natural gas for the West European continent until well into the 1990s, and in any case a plan needed to be worked out to finance such developments in the absence of attractive commercial incentives. It is highly probable that deliveries of increased quantities of both U.S. coal and Norwegian natural gas will play an important part in West European energy supplies in due course but not as direct replacements for Soviet natural gas. On the other hand, increased deliveries of Soviet natural gas are unlikely to preclude the necessity for West European imports of U.S. coal or Norwegian gas, and it should be borne in mind that even though volumes of natural gas supplied by the Soviet Union now seem slated to increase from 23 mrd m^3 in 1980 to nearer 60 mrd m^3 by the early 1990s, imports of Soviet crude oil and products into Western Europe that amounted to some 50 to 55 million tons per annum in the early 1980s seem destined to decline substantially during the decade as Soviet export surpluses dwindle under the triple pressures of growing domestic demand, East European demand, and a leveling off of production.

PROSPECTS FOR THE FUTURE

The question of West European energy-import dependency will remain a difficult one for the foreseeable future, and the issue of the natural gas dependency of some West European countries upon the Soviet Union is not likely to disappear. However, there is every reason to expect that the relevant West European countries will be in a better position than ever to manage their energy and security relationships with the Soviet Union as a result of the impetus that the United States has given to the matter, and it is to be hoped that a contribution has thus been made toward the formulation of an even more coherent and comprehensive Western European energy policy than may have hitherto existed. It is also to be hoped that those who are in principle opposed to the expansion of East-West trading relationships in general and to the realization of new natural gas export projects in particular will carefully evaluate not only the damage that they believe will be done to the Soviet economy if the projects are killed but also that which will be done to both the West European economies and to the Western Alliance, perhaps less tangibly but nonetheless effectively. The inconsistency with which they have based their opposition upon either economic, security, or political grounds suggests that their real reasons may have been more ideological than anything else. In view of the intensity with which the Soviet Union concentrates upon the ideological confrontation with the West, there is clearly a need for the West to look to its ideological defenses and to mobilize these strengths to ensure its continued survival. However, the selection of the Yamburg-Urengoi Project as a major weapon in the ideological struggle without having secured the approval and support of those Western nations most intimately involved in it would appear to have risked weakening, not strengthening, the West's ideological cohesion. This would be especially serious if the transatlantic differences of opinion over the issue, which appeared in mid-1982 to be deepening, remained unresolved while the project itself or, at any rate, the practical outcome of increased Soviet natural gas exports to the West, seemed eventually likely to be realized in one way or another. Was the attempted killing of the project really the best answer the West could come up with in the East-West confrontation? And was not perseverance with such a policy threatening to produce a net gain in points for the Soviet Union in the long run? It is difficult to avoid the conclusion that the Yamburg-Urengoi Project was the wrong issue—chosen at the wrong time and handled in the wrong way—on which to make an ideological stand and that the incident was given far more prominence in international affairs than it merited.

The issue of Soviet natural gas exports to Western Europe is not going to disappear with the coming on stream of Yamburg-Urengoi deliveries in the mid-1980s. Some compromise solutions to various problems arising from divergent transatlantic approaches to the Yamburg-Urengoi Project were in the course of being found as 1982 drew to a close. These would enable U.S. sanctions to be lifted, in exchange for a somewhat firmer and better coordinated Western attitude to the extension of credits and the selling of technology to the Soviet Union and Eastern Europe. However, much still needs to be done to straighten out the framework within which East-West trade is to be conducted in future if further damaging dissent between Western allies is not to be generated by new large-scale Soviet natural gas proposals, which seem almost certain to be put forward from time to time over the next few decades.

9 DISCUSSION AND CONCLUSIONS

One never really comes to the end of a journey such as the one embarked upon at the beginning of this book. Every step of the investigation suggests new angles and new avenues to explore, and the actors change their styles even as the scenes shift upon the revolving stage. Nevertheless, perhaps the investigation has revealed enough to indicate that the role natural gas seems likely to play in meeting Western Europe's energy requirements over the next two decades is, for reasons it is hoped are by now clearer, liable to be different from the role that it could play. What role it actually will play is determined by many factors that are not susceptible to measurement or prediction. The only one that can perhaps be taken for granted is uncertainty itself, particularly uncertainty about demand and supply.

Faced with such a situation, potential investors in the international natural gas industry might be forgiven if they turned their attention elsewhere, but they would probably find similar conditions prevailing in many other sectors of international trade—particularly those related to the mobilization of the world's energy resources. But Western Europe is going to have to import hydrocarbons for several decades to come, and while it may continue to be successful in reducing its oil imports, there seems little chance that, even if economic growth and energy consumption become further decoupled as a result of conservation and greater end-use efficiency, imports of natural gas will do other than steadily increase in view of the declining production of indigenous resources and the time it would take to bring some of the promising new discoveries in the North Sea on stream. The infrastructure for distributing and using gas is already in place and, particularly in the residential sector of the market, it would be a long and costly process to replace natural gas with some other energy form even if one were readily available that did not require imports and was not therefore subject to the vagaries of international trade. Electricity from nuclear reactors may well compete with natural gas for the residential space-heating markets one day, provided more governments can resolve the problems, arising both from the technical choices open to them and from the antinuclear groups, as successfully as the French appear to have solved theirs. But it seemed unlikely, in late 1982, that nuclear-produced electricity would emerge as a serious competitor for natural gas outside France until the end of this century at the earliest. For practical reasons if for no other, the burning of coal in the home hearths of millions of apartment blocks constructed over the past twenty years with neither chimneys nor storage bunkers can be ruled out.

So natural gas is going to continue to be used in Western Europe for a very long time, and, if it is physically available, there is ample scope for it to increase its share of the energy markets in many countries, provided the price is right. Indeed, as we have seen, natural gas is forecast to play a somewhat greater role in meeting the energy requirements not only of relative newcomers to the scene—such as Spain, Denmark, and Sweden—but also of some of the established gas users, such as West Germany. In the early 1980s, schemes for the supply of gas in the 1990s and beyond have to be created and set in motion. The buyers have to go searching for sellers in different countries, and producer countries have to decide how much of their gas they want or need to sell abroad and which consumer countries offer them the most profitable and secure long-term markets. In spite of all the geopolitical clouds on the horizon and lions in the path, new deals will be done between various West European countries and various traditional or new producer countries. They have to be done, and it may be that people one day will come to wonder how so much attention came to be paid in the late 1970s and early 1980s to geopolitical factors affecting the future of Western Europe's gas supplies.

It appears that the heightened interest may have stemmed from two principal sources: the developing tension between the two superpowers and the insecurity felt by the oil-importing countries of the

West as a result of the 1973 and 1979 developments in oil markets (and the Middle East generally) and their implications for the future. It was not surprising that some people in the West should have associated the energy problems apparently being faced by the Soviet Union, according to a CIA report published in 1977, with the Soviet invasion of Afghanistan in 1979. There were two main lines of reasoning. The first suggested that, since the Soviet Union was likely to become a net oil importer by the late 1980s and the COMECON bloc was likely to become a substantial overall importer by the mid-1980s, it would need to secure access to overseas oil supplies for itself and its allies by fair means or foul. Furthermore, since the COMECON countries were chronically short of hard currency, and the oil-exporting countries of OPEC had, at the time, all the petrodollars they knew what to do with, there seemed little that the former would be able to offer the latter as an inducement to part with its oil—nothing, that is, except possibly military hardware, and there are limits to the amount of that that a country can absorb even if it is of the highest quality and there are no political strings attached. If the COMECON countries would not be able to pay for their oil like "normal" countries, so the reasoning went, they would have to secure it by other means, such as by overt force or by bringing about the collapse of Middle Eastern regimes that adhered to the capitalist camp and their replacement by progressive regimes willing to assist the Soviet cause by providing oil on favorable terms. The invasion of Afghanistan was seen by some as a step in this direction, aimed at encouraging a more accommodating attitude on the part of the oil exporters toward the Soviet bloc. The other line of reasoning recalled the concept of the "Black Crescent," (black being the color of crude oil) stretching from North Africa through the Middle East and up into Siberia, in which some two-thirds of the world's hydrocarbon resources are probably located. If, according to this argument, one could control that crescent one could gain powerful influence over the allegiances of all hydrocarbon-importing nations. Since the Soviet Union already encompasses about half of the crescent, it need not actually conquer the remainder militarily. It could simply create such instability there that the access to the hydrocarbons that the developed countries of the West had enjoyed for so long and upon which they had become so dependent would be interrupted, and a major source of their economic strength would be denied them. Once again, the Soviet invasion of Afghanistan did nothing to allay the suspicions of those who doubted Soviet intentions even though they were reluctant to condemn them outright, but it greatly strengthened the hand of those who firmly believed that the Soviet Union was out to disadvantage the West by any means possible, short of war. In mid-1982, the Soviet Union appeared likely, in spite of a surprising degree of resistance shown by the Afghan people, to advance one step nearer to the Arabian Gulf—a better position from which to influence affairs in the gulf and probably also in Iran, where the political situation had suddenly become very unpredictable with the overthrow of the Shah and the war with Iraq.

Those in the West who were, understandably, concerned by happenings in the Middle East at the end of the 1970s but who saw the hand of the Soviet Union behind every adverse development there, tended to be less sensitive than perhaps they should have been to some of the other causes of instability in the area and to the feelings, hopes, and fears of the people who actually lived there. For example, few Arabs saw the Soviet Union as the major threat to world peace or, indeed, as the major cause of instability in their part of the world. Instead, they regarded Israel as by far the most disruptive element, and they bitterly resented Western support for a country that was their sworn enemy and that was in their eyes the arch aggressor. So much did the Arabs resent Western support for Israel that some of them, the "front-line" states like Syria and Libya, were prepared to accept even Soviet military and political assistance in their fight against the Israelis, although they were not friends of communism as such and were apprehensive about encouraging too close an involvement of another superpower that they had no cause to love. The Soviet Union, though obviously a regional power, is not "of the region," and its ideology is incompatible with the teachings of the Koran. It does, moreover, have a largely Moslem population in its central Asian republics and must be extremely concerned that the manifestations of Islamic fundamentalism that it observes in Iran may have unpleasant repercussions for itself one day. But so long as the Arab-Israeli dispute exists, the Soviet Union and some other countries will consider that they have a role to play in the region, quite apart from any commercial interest they might have in securing access to Middle East oil.

In 1982, although the performance of the Soviet coal industry was still falling far short of expectations and that of the oil industry could only be described as lacklustre, the state of the Soviet energy supply situation overall was not as bad as some in the West had been forecasting, and it did not seem to indicate that the Soviet Union would be left with no alternative but to attempt to lay its hands upon the oil wealth of the Middle East. Furthermore, if it and its allies could somehow find ways acceptable to the oil-producing countries to purchase increased supplies of their oil, then there appeared little reason, in the early 1980s, why the Soviet Union should not be perfectly at liberty to do so. However, the Soviet Union's experiences in trying to bring the Afghanistanis under control and the almost universal condemnation—not only by its ideological opponents in the West but also by the nonaligned countries in general and by the Moslem nations in particular—with which its actions in that country were greeted must have thrown considerable doubt upon its actual ability to achieve much of lasting value in the Middle East by force. Furthermore, following the invasion of Afghanistan, the United States had made it very clear that it was not prepared to tolerate further attempts at Soviet expansionism in this area of vital interest to the West.

The Iran-Iraq war that started in late 1980 was still continuing a year and a half later, when the Arab countries found themselves divided in their support for the combatants, with Syria and other radical groups anxious to see the downfall of the Iraqi regime for largely factional reasons, and both Saudi Arabia and Egypt anxious that the religious fanatics in Iran did not gain a military victory that might greatly strengthen the cause of the fundamentalism which, in turn, threatened the very existence of the Saudi, Egyptian, Kuwaiti, Jordanian, and other Arab regimes. The Soviet Union needed to tread extremely carefully amid all this turmoil, even though some in the West have suggested that, whatever the outcome of the Iran-Iraq war, the Soviet Union will tend to benefit from the instability that has ensued and that it will be in a strong position to cooperate with or influence that leadership which emerges in Iran following the passing of Khomeini. It is probable that it will find it expedient to act in the direction of preserving rather than destroying stability on its southern flank, at least until such time as it is in a position to control events there—something that it had not shown in any great capability of doing by the middle of 1982.[1]

1. J.B. Hannigan and C.H. McMillan, *The Soviet Energy Stake in Afghanistan and Iran: Rationale and Risk of National Gas Imports*, Research Report #16 (Ottawa, Canada: Carleton University, 1981).

Many in the Middle East would prefer that neither the Soviet Union nor the United States be the predominant power there and would even rejoice if neither were heavily involved except strictly by invitation. To this end they created in the early 1980s a Council for Cooperation on security matters in the Arabian Gulf, and while it seemed unlikely that it would ever evolve to the point where the combined military forces of the member countries would be able to withstand aggression on a large scale, its creation may have been a signal to both the West and the East that there are influential forces there who wish increasingly to shape their own futures. If only the accident of oil under their territories did not bring with it so many headaches and responsibilities of an international nature, the Arab countries would have a much easier time of it; many have, nonetheless, exhibited a rare degree of statesmanship in handling the complex situation in which they find themselves.

The purpose of focusing for a moment upon the Middle East has been to emphasize that around this one area were concentrated many of the main geopolitical issues affecting East-West relations, at least in the eyes of many in the United States in the early 1980s, and that at the heart of the matter was oil and future access to it. So prominent a part was the Middle East playing in international affairs at that time that it tended to distract attention from what was happening elsewhere and to suggest that aspects of geopolitics other than the East-West confrontation, such as growing nationalism and regionalization, were of minor importance. But there are those who have suggested that, in view of its enormous reserves of natural gas, the Soviet Union could one day come to play the same pivotal role in respect of global supplies of natural gas that the Middle East has played in respect of oil, although the contrast between the publicly controlled energy industries of the Soviet State and the fragmented history of the many countries and actors in the Middle East oil saga—which allowed for, among other things, the active participation of international investment capital and private enterprise—would indicate that such is unlikely to be the case. There is little doubt, however, that Soviet natural gas is going to play an increasing role in Western Europe's gas industry, and although West Europeans are aware of the potential dangers of this situation, they are also aware both of the benefits to be derived from it and of what they must do to ensure that that role does not become too important or an unmanageable geopolitical millstone around their necks.

The international gas business is, if the regionalization that seems to be developing in the international order as a whole is anything to go by, also likely to develop along regional lines, and while there may be exceptions, it is likely that Western Europe will actually come to take the bulk of its imported gas from those countries closest to it, just as Japan will rely largely upon Pacific Basin suppliers and the United States will rely upon Canadian and Mexican imports. The interregional environment is likely to remain so unstable for the rest of the twentieth century as to militate against significant exchanges of what is, after all, a considerably less fungible form of energy than crude oil. If there were no problems over capital formation, many longer distance schemes, particularly for liquefied natural gas (LNG), might be attempted on logical logistical grounds, and gas from the Middle East and West Africa might find itself flowing through pipelines across North Africa or the Balkans on the way to Western Europe. But, in a climate of high real interest rates, it is difficult to attract investors to schemes where there is little guarantee of security of supply and where even the size of demand over two decades is in question. Investors will prefer to concentrate on developing energy resources where there is the greatest possible degree of control in the "right" hands. The problem then arises as to how to judge the "rightness" of hands, which leads into the field of political risk analysis.

The relevance of political risk analysis for those attempting to assess the desirability of a particular country as a potential partner in a multibillion-dollar quarter-century deal may not be so obvious in the natural gas industry as it may be in other cases. Few regimes these days can confidently expect to remain in power for ten, let alone twenty-five, years, and there is certainly no guarantee that even if they did they would adhere to the same fiscal and resources-development policies throughout their tenure in office. West Europeans are well aware of how, even under what they might regard as exemplary governmental and political conditions, changes in democratically elected regimes can result in profound changes in policies and attitudes toward business, finance, and raw-materials development. In some other parts of the world, regimes change for a number of different reasons, some of which have little to do with the will of the people, and in weaker countries, regimes are increasingly subject to geopolitical manipulation by powers concerned primarily with furthering their own objectives. In such circumstances, it seems more important to try to identify those cases in which the long-term interests of a country are going to be best served by involvement in a natural gas export scheme, whatever the political color of the regime in power, than attempt to quantify the probability that a given regime will change within a given period of time. Political risk analysis has not been very successful in predicting regime changes, and it also gives little guidance about what these will actually mean for the business environment in general, let alone any particular sector of it.

The gas-consuming countries of Western Europe are, between them, in a unique position, as a result of their colonial experiences, to understand the needs and aspirations of those countries that could be considered potential natural gas suppliers over the next two decades, and this should greatly facilitate the putting together of mutually beneficial projects. Western Europe is indeed fortunate that there are so many deposits of exportable natural gas located within a reasonable distance for both pipeline and LNG supplies to be feasible. Looked at another way, however, it is fortunate for many potential natural gas exporting countries that the large West European market exists relatively nearby and that a number of options are available for producer and consumer countries to benefit from each others' presence. It would seem logical, therefore, for the number of natural gas schemes to proliferate during the next two decades, provided that countries' perceptions of their own and others' actual abilities and requirements approximate the changing realities and provided that nobody is too greedy. What could well prevent more gas from flowing would be the adoption by either the buyers or the sellers of unrealistic positions over the amount of benefit they should receive. If, for example, West European buyers should seek supplies at prices that are too low, producer countries will be encouraged to develop schemes to utilize their gas domestically or to keep it in the ground. If the producers should hold out for prices that are too high, the West European natural gas market will cease to expand, and alternative energies will be used. The adoption of extreme positions can surely be avoided by the development of genuine communication and understanding between producers and consumers.

It is the availability and pricing of alternative fuels, added to the fact that natural gas only accounts for some 15 to 17 percent of total energy consumption in the main West European markets, that reduces the potential for geopolitical manipulations to influence the future of Western Europe's natural gas supplies. Some potential supplies of natural gas will obviously be preferred by West European buyers over others, and not all of them will be required at the same time for the same markets. In the same way that there is no global market price for natural gas, so there is no minimum amount of money that all producers should expect to net back to their wellheads when they export their gas. A country like Indonesia, with a vigorously growing population and ambitious economic development plans, may be happy to sell as much gas as it can to Japan or to any other country that will buy it at a reasonable price. Algeria, on the other hand, with its substantial, albeit declining, earnings from oil exports and a slower growing population, may consider that it can afford to shepherd its natural gas resources until such time as a buyer appears who will pay a premium price.

Compromise is the key to the future of the West European natural gas supply situation. Also important is flexibility, which allows the creation of contract terms with sufficient leeway to accommodate unforeseen major fluctuations in business conditions and discourages both sides from either adopting rigid negotiating positions, which lead nowhere, or from attempting unilaterally to impose postcontract changes in agreed terms. There obviously should be a justifiable premium for the depletion of a country's finite reserves of whatever raw material, but it is far from established that this premium should be the same all over the world. Indeed, it cannot be so. But there is considerable onus on those who need or desire to consume somebody else's finite resource to provide a sufficient incentive for the producer country to make the effort to mobilize that resource. In the case of natural gas, where such a long-term partnership has to be entered into well before any export project can be begun, both sides must feel they have an equal stake in the successful outcome of that project and that no one partner is getting an unfair share of the benefits or shouldering an unfair share of the investment burden.

While economies of scale can be extremely important in the natural gas business, it may turn out that the optimum size for, say, an LNG plant or export terminal is just too large for an exporting country to afford. Many potential exporters may face considerable problems in borrowing enough money to make the appropriate investment in their end of a project, and there is often something to be said for both partners adopting a more gradual approach. There are always enemies of any project—particularly one that, for example, involves a small or economically and militarily weak exporting country on the one side and a rich and powerful consumer nation on the other. If escalating financial problems are added to the burden of involvement in neocolonialism or economic neoimperialism then opposition parties may well attract sufficient popular support either to destabilize a regime or sabotage a project—both liable to lead to considerable delays in getting it off the ground. If the World Bank or some such august and powerful financial body can be persuaded to become involved, the very fact that it is associated with a project can strengthen the positions of both parties vis-à-vis their own people and give each confidence that neither is being taken advantage of.

West European natural gas importers seemed, in 1982, to be well placed to meet market needs until at least the end of the decade, and in some cases well into the 1990s, provided that all the schemes under contract or advanced negotiation came to fruition. Closer examination of the situation reveals that in some cases, however, there were project delays in store caused by developments in producer or potential producer countries, while in other cases there appeared to be the prospect of oversupply resulting from downgraded forward estimates of demand. It is toward the reconciliations of these two positions that people involved with all aspects of the discovery, production, and distribution of natural gas have to devote their attentions, although it is virtually impossible to achieve a perfect match between supply and demand. The framework within which they have to operate is less flexible than that of the oil industry, and there are no funds available for creating the degree of redundancy that would lead to the creation of, for example, a spot market for natural gas. This would involve numerous loading or landing terminals significantly underutilized and a number of extremely expensive, fully laden LNG tankers loose on the high seas looking for someone able and willing to buy their cargoes and to store them until required for use. The West European utilities have to make sure that the major proportion of their customers' requirements are firmly covered—since they only have a finite number of interruptible or bivalent end users who could cushion the blow of a supply interruption caused by either project failure or a deliberate political move on the part of a producer country—but they must at the same time be careful not to put all their energy eggs into one supply basket, particularly if that basket lies outside the boundaries or physical control of the West Europeans themselves.

The way that the supply of imported natural gas to Western Europe has evolved over the past two decades demonstrates the cautious approach that gas people adopt to the question. Not all the available gas from any one source has been mobilized to the exclusion of other sources, and if projects under discussion in the early 1980s are successfully brought on stream, the diversification of Western Europe's supplies will even improve by the year 2000, when Nigeria, Cameroon, and Qatar are added to the list. The three basic suppliers for the major continental West European markets will remain the Soviet Union, Norway, and Algeria for the forseeable future, and supplies from the Soviet Union will assuredly assume a larger proportion of imported gas as a result of developments already in the pipeline. But those who objected to the latest Soviet export development had not, by the autumn of 1982, convinced many people in Western Europe that this was folly on the part of West European buyers. The methods of financing the acquisition of additional volumes of Soviet gas may have been more appropriately criticized, but the volumes themselves and the proportions that they would come to represent in various countries' fuels/energy balances could hardly be termed excessive or dangerous, given the comparatively lowly role that natural gas plays in West European overall energy consumption.

Western Europe remains faced with the twin problem of diversifying its energy imports away from oil and away from the Middle East. Solutions to this problem will not occur overnight, and it will indeed be well into the next century before energy sources other than hydrocarbons and coal will come to play a significant part in meeting West Europe's requirements. Looking at things over the longer term, it could turn out that Soviet exports of gas to the West peak by the end of the century as increasing quantities are required domestically and as political pressures arising out of security considerations militate against the construction of yet further Soviet export projects. But if it is accepted that natural gas is not such a strategically important commodity as oil and that the exporters in any international natural gas arrangement have a great deal to lose if supplies are interrupted, particularly for political reasons, there seems no good reason why the West Europeans should not continue to alleviate their energy supply-source diversification problems with the aid of Soviet gas, so long as it is available and offered at an acceptable price. Its availability certainly strengthens the West Europeans in their negotiations with, for example, the Algerians and other potential suppliers.

It would probably benefit neither side if Soviet gas did come to play a much larger role in Western Europe's energy supply situation, even though this would have been most unlikely to happen even before the hue and cry over the Yamburg-Urengoi Project. No one has yet attempted to define what is or is not an acceptable level of imports from a given country, particularly if that country is regarded in many quarters as being inherently hostile, but there is little to be gained from attributing a greater geopolitical danger than can be satisfactorily demonstrated to a commercially and logistically attractive transaction—especially one involving a commodity that is rather unsatisfactory as a potential geopolitical weapon. Contingency planning should enable West Europeans to design for the problems that could arise from increased purchases of Soviet gas and to minimize the risks involved. International trade in natural gas involves a high degree of compromise and commitment on the part of all involved—not a bad basis, one would think, upon which to attempt to create a slightly less dangerous world. It would be totally illogical if Western Europe, with 15 percent of the world's market for natural gas, were to ignore the existence of Soviet offers from proven reserves representing over 35 percent of global resources, persisting instead in trying to secure supplies from areas of the world either no better renowned for their stability and reliability as supply sources or afflicted with geographical or technological problems such as to severely limit their attractiveness or significantly postpone their accessibility.

The Middle East, North Africa, and the Soviet Union might not be everybody's first choices for ongoing hydrocarbon supply sources, but they are the only practical possibilities for Western Europe at the beginning of the great transition period that the world is facing. Those who insist that all the billions of dollars on their way to the Soviet Union are functioning to pay for the next strand in the rope with which the Communists will ultimately hang the capitalists, and would better be spent on developing the North Sea gas deposits or importing vastly increased amounts of coal, are either unaware of the technological and other practical problems involved, given the time frame of West European energy estimates, or have not thought through the difficulties of bringing about commercially feasible solutions to essentially political problems. No imported energy scheme should be expected to be entirely satisfactory from the point of view of security of supply, but there must surely be greater strength in greater numbers, and it would be just as irresponsible if the West Europeans were to rely only upon North Africa or the Middle East for their gas imports as it would be if they chose to take only Soviet gas. It will be a long and painful process diversifying away from imported hydrocarbons and from the Middle East, but it has to be done. It is to be hoped that those who have been concerned about the geopolitical implications for future natural gas supplies for Western Europe will have fewer doubts—or grounds for misunderstandings should major export projects—whether from the Soviet Union or from some other country—crop up in the future. This would occur partly as a result of the airings that the issue has had over both the Algerian and the Soviet experiences and partly as a result of the journey on which this book has taken them.

In conclusion, it appears that although geopolitical factors will continue to influence the development of the international trade in natural gas and give rise to limitations that, at the time of writing, seemed likely to restrict the number of possibilities for supplying Western Europe's future import needs, it does not appear as though natural gas can be successfully used as an instrument of geopolitical manipulation. This is largely because of the relatively minor role in countries' overall fuels and energy balances that natural gas seems destined to play, given the availability of alternatives and the infrastructural problems associated with its transportation and, in particular, the penalty of distance. For the remainder of the 1980s, Western Europe will be faced more with a demand than a supply problem for natural gas, but every effort should now be being made to achieve the degree of mutual understanding between producer and consumer countries that will enable this extremely high-quality energy source to maintain and expand its contribution to meeting West Europe's energy requirements in the 1990s and well into the twenty-first century.

BIBLIOGRAPHY

The following selection of sources, in addition to such essential reading as the Petroleum Times, the Petroleum Intelligence Weekly, and the Geopolitics of Energy, will be found particularly useful to those wishing to pursue their own investigations. Many of the sources have significance for more than one sector, but they are listed under the chapters for which they have the greatest relevance.

CHAPTER 2

Deese, David A. and Joseph S. Nye. *Energy and Security.* Cambridge, Mass.: Ballinger Publishing Company, 1981.

International Institute for Applied Systems Analysis. *Energy in a Finite World.* Paths to a Sustainable Future, vol. 1, and A Global Systems Analysis, vol. 2. Cambridge, Mass.: Ballinger Publishing Company, 1981.

Wagner, G.A. *Energy: The Double Dilemma.* Address to the International Chamber of Commerce, Swedish National Committee, Stockholm, March 25, 1981.

CHAPTER 3

Goff, Rene Le. *Natural Gas and other Gases, The Economics of Gas.* Paris: Claude Detourne, Total Information # 87, 1981.

Jensen, James T. *World Natural Gas Reserves and the Potential for Gas Trade.* Middle East Economic Survey, Supplement. January 28, 1980.

Johnson, Daniel W. "Gas Supply and Demand, Global View: Long and Short Term." Paper presented at The Bahrain Society of Engineers Seminar on the Future Role of Natural Gas in Meeting Energy Needs, Bahrain, 14 and 15, November 1981.

Lichtblau, John H. *U.S. Natural Gas Policy and Outlook for Gas Imports.* New York: Petroleum Industry Research Foundations, Inc., 1982.

Linden, Henry R. *U.S. Energy Outlook—1981.* Chicago: Gas Research Institute, October 1981.

Natural Gas—BP Briefing Paper, London: British Petroleum, September 1981.

Peebles, Malcolm W.H. *Evolution of the Gas Industry.* London: Macmillan, 1980.

CHAPTER 4

Jarjour, Gabri. "Long-Term Gas Pricing Strategy." *Journal of Energy and Development* 7, no. 1. (Autumn 1981): 89–98.

CHAPTER 5

Belguedj, D.M. "The Supplies of Natural Gas in Western Europe." Paper presented at the European Gas Conference, Oslo, 25 May 1981.

Frazer, Frank. "Gas Prospects in Western Europe," *The Financial Times,* London: Business Information Limited, 1981.

Greer, Boyce I. and Jeremy L. Russell. *European Reliance on Soviet Gas Exports: The Yamburg-Urengoi Natural Gas Project.* Harvard Energy Security Program, Discussion Paper, John F. Kennedy School of Government, February 1982.

Hastings, Earl A. *Energy Challenges Facing the Atlantic Alliance.* Report to Economic Committee of the North Atlantic Assembly, International Secretariat, October 1981.

Maull, Hanns W. "Natural Gas and Economic Security." *The Atlantic Papers,* No. 43. Paris: The Atlantic Institute for International Affairs, 1981.

Ouwehand, Floris. "The Role of Natural Gas in Western Europe in the Next 20 Years." Paper presented at the European Gas Conference, Oslo, 25 May 1981.

CHAPTER 6

Aliboni, Roberto. *Libys's Inter-Arab and Regional Politics.* Rome: Instituto Affari Internatzionali, 1981.

The Atlantic Council of the United States. *U.S. Policy Towards Canada: The Neighbours we Cannot Take for Granted.* Washington, D.C.: The Atlantic Council, 1981.

Conant, Melvin A. and Eugene Maechling. *Political Aspects of A New World Energy System.* Washington, D.C.: Georgetown Center for Strategic and International Studies, 1980.

"Documentation on Algerian Energy Policy." *Middle East Economic Survey,* January 12, 1981.

CHAPTER 7

Bull, Hedley. *The Anarchical Society.* London: Macmillan, 1977.

Keohane, Robert O. and Joseph S. Nye. *Power and Independence.* Boston: Little, Brown, 1977.

Martin, Laurence. *The Two-Edged Sword.* London: Weidenfeld and Nicholson, 1982.

Morgenthau, Hans J. *Politics Among Nations.* New York: Knopf, 1948, revised 1978.

Stanford Research Institute. *Values and Strategic Planning.* SRI International "Values and Lifestyles" Report No. 15, November, 1980.

CHAPTER 8

Bethkenhagen, Jochen. "Oil: Resources, Exploration and Projected Production through 1990," in *CMEA: Energy 1980-1990.* Brussels: Nato Economics Directorate, 1981.

Cobb, Tyrus W. "East-West Energy Trade and Technology Transfer: A Security Perspective," *CMEA: Energy 1980-1990.* Brussels: Nato Economics Directorate, 1981.

Dienes, Leslie and Theodore Shabad. *The Soviet Energy System.* Washington, D.C.: V.H. Winston, 1979.

Energy in Soviet Policy, a Study for the Joint Economic Committee. Washington, D.C.: Government Printing Office, 1981.

Goldman, Marshall I. *The Enigma of Soviet Petroleum.* London: George Allen and Unwin, 1980.

Gustafson, Thane. *Selling the Russians the Rope? Soviet Technology Policy and U.S. Export Controls.* Santa Monica, Calif.: The Rand Corporation, April, 1981.

Hannigan, J.B. and C.H. McMillan. *The Energy Factor in Soviet-East European Relations.* East-West Commercial Relations Series No. 18. Ottowa: Carleton University, 1981.

"Joint Energy Projects and their Influence on Future COMECON Energy Autarchy Ambitions," *COMECON: Progress and Prospects.* Brussels: Nato Directorate of Economic Affairs, 1979.

Minard, Lawrence. "Embracing the Bear. *Forbes Magazine,* June 7, 1982.

Muller, Friedmann. "Implications for the West, Including Defense, of CMEA Energy Prospects," *CMEA: Energy 1980–1990.* Brussels: Nato Economics Directorate, 1981.

National Foreign Assessment Center. *USSR: Development of the Gas Industry.* Washington, D.C., 1978.

Russell, Jeremy L. *Energy as a Factor in Soviet Foreign Policy.* London: Saxon House/Royal Institute of International Affairs, 1976.

Stern, Jonathan P. *Soviet Natural Gas Development to 1990.* Lexington, Mass.: D.C. Heath, 1980.

——*East European Energy and East-West Trade in Energy.* Energy Paper No. 1, British Institutes' Joint Energy Policy Programme, Policy Studies Institute. London: Thetford Press Limited, 1982.

"The U.S. Acts to Plug Russia's Credit Line." *Business Week,* March 22, 1982, p. 27.

U.S. Congress, Office of Technology Assessment. *Technology and Soviet Energy Availability.* Washington, D.C.: Government Printing Office, 1981.

"Will Russia's Pipeline to Europe Stop in Washington?" *The Economist,* American Survey, March 6, 1982, p. 19.

CHAPTER 9

Hannigan, J.B. and C.H. McMillan. *The Soviet Energy Stake in Afghanistan and Iran: Rationale and Risk of Natural Gas Imports.* East-West Commercial Relations Series No. 16. Ottawa: Carleton University, 1981.

I.E.A. *Natural Gas: Prospects to 2000.* Paris: I.E.A. Secretariat, 1982.

International Communication Agency, Office of Research. *Soviet Elites: World View and Perceptions of the U.S.* International Communication Agency, September, 1981.

INDEX

ABOUT THE AUTHOR

Jeremy Russell received a master's degree in Zoology at Pembroke College, Cambridge University. In 1959 he joined a major international chemical company based in London. His knowledge of Russian led him to become involved in trading with the Soviet Union and other Eastern European countries, and he has traveled extensively in that part of the world. He has studied the energy situation in Eastern Europe, spent a sabbatical year as Research Fellow at the Royal Institute of International Affairs, Chatham House, in London, and has spent an academic year as a Visiting Research Fellow at the John F. Kennedy School of Government, Harvard University. His first book, *Energy as a Factor in Soviet Foreign Policy,* was published in 1976. He has published articles in *The World Today, The Times* (London), *The World Economy,* and *The Energy Journal.*